Springer Theses

Recognizing Outstanding Ph.D. Research

Aims and Scope

The series "Springer Theses" brings together a selection of the very best Ph.D. theses from around the world and across the physical sciences. Nominated and endorsed by two recognized specialists, each published volume has been selected for its scientific excellence and the high impact of its contents for the pertinent field of research. For greater accessibility to non-specialists, the published versions include an extended introduction, as well as a foreword by the student's supervisor explaining the special relevance of the work for the field. As a whole, the series will provide a valuable resource both for newcomers to the research fields described, and for other scientists seeking detailed background information on special questions. Finally, it provides an accredited documentation of the valuable contributions made by today's younger generation of scientists.

Theses are accepted into the series by invited nomination only and must fulfill all of the following criteria

- They must be written in good English.
- The topic should fall within the confines of Chemistry, Physics, Earth Sciences, Engineering and related interdisciplinary fields such as Materials, Nanoscience, Chemical Engineering, Complex Systems and Biophysics.
- The work reported in the thesis must represent a significant scientific advance.
- If the thesis includes previously published material, permission to reproduce this must be gained from the respective copyright holder.
- They must have been examined and passed during the 12 months prior to nomination.
- Each thesis should include a foreword by the supervisor outlining the significance of its content.
- The theses should have a clearly defined structure including an introduction accessible to scientists not expert in that particular field.

More information about this series at http://www.springer.com/series/8790

John S. Van Dyke

Electronic and Magnetic Excitations in Correlated and Topological Materials

Doctoral Thesis accepted by University of Illinois at Chicago, Chicago, Illinois, USA

John S. Van Dyke
Department of Physics and Astronomy
Iowa State University
Ames, Iowa, USA

ISSN 2190-5053   ISSN 2190-5061 (electronic)
Springer Theses
ISBN 978-3-319-89937-4   ISBN 978-3-319-89938-1 (eBook)
https://doi.org/10.1007/978-3-319-89938-1

Library of Congress Control Number: 2018942180

© Springer International Publishing AG, part of Springer Nature 2018
This work is subject to copyright. All rights are reserved by the Publisher, whether the whole or part of the material is concerned, specifically the rights of translation, reprinting, reuse of illustrations, recitation, broadcasting, reproduction on microfilms or in any other physical way, and transmission or information storage and retrieval, electronic adaptation, computer software, or by similar or dissimilar methodology now known or hereafter developed.
The use of general descriptive names, registered names, trademarks, service marks, etc. in this publication does not imply, even in the absence of a specific statement, that such names are exempt from the relevant protective laws and regulations and therefore free for general use.
The publisher, the authors and the editors are safe to assume that the advice and information in this book are believed to be true and accurate at the date of publication. Neither the publisher nor the authors or the editors give a warranty, express or implied, with respect to the material contained herein or for any errors or omissions that may have been made. The publisher remains neutral with regard to jurisdictional claims in published maps and institutional affiliations.

Printed on acid-free paper

This Springer imprint is published by the registered company Springer International Publishing AG part of Springer Nature.
The registered company address is: Gewerbestrasse 11, 6330 Cham, Switzerland

For my wife, Esther

Supervisor's Foreword

Understanding the complex properties of strongly correlated electron materials has been an outstanding problem at the forefront of research in condensed matter physics for nearly 40 years. It was stimulated by the discovery of the heavy fermion superconductors and the quest for identifying the microscopic mechanism responsible for the emergence of their unconventional superconducting phase. The similarity of the heavy fermion's phase diagram with that of subsequently discovered unconventional superconductors, such as the cuprate (high-temperature) or iron-based superconductors, has raised the question of a common, universal pairing mechanism. In particular, the proximity of the unconventional superconducting phase to antiferromagnetism in the phase diagram of all of these materials has given rise to the hypothesis of a pairing mechanism mediated by the exchange of antiferromagnetic fluctuations. For the heavy fermion materials, whose salient feature is a lattice of magnetic moments that are either Kondo screened by conduction electrons or ordered antiferromagnetically, this hypothesis has remained unproven despite an impressive body of theoretical and experimental studies. A major obstacle in verifying the hypothesis has been a lack of insight into the complex electronic and magnetic structure of these materials.

The work by Dr. John Van Dyke described in this book represents a major breakthrough in exploring and confirming this 30-year-old hypothesis for the heavy fermion material $CeCoIn_5$, considered to be a prototype material for the entire class of heavy fermion compounds. Dr. Van Dyke demonstrated—making use of recent groundbreaking quasiparticle interference (QPI) experiments by the group of Prof. J.C. Seamus Davis (Cornell University)—that characteristic signatures in the QPI spectrum of $CeCoIn_5$ can be employed to extract not only the momentum form of its superconducting order parameter—exposing its unconventional $d_{x^2-y^2}$-symmetry—but also the multi-band electronic structure crucial for the emergence of superconductivity. However, to quantitatively identify the microscopic pairing mechanism, a second crucial, and so far missing, element was necessary—the form of the superconducting pairing interaction that was proposed to arise from the antiferromagnetic coupling between the localized moments. Dr. Van Dyke showed that the momentum structure of this interaction can be extracted from the experimental QPI data, allowing him to develop a quantitative microscopic theory for the unconventional superconducting state in $CeCoIn_5$. This work resulted in seven predictions for this material's striking physical properties: the symmetry and momentum structure of the multi-band superconducting order parameter, the critical temperature, the momentum and energy dependence of the QPI as well as the phase-sensitive QPI spectrum, the temperature dependence of the spin-lattice relaxation rate, the energy position of the magnetic resonance peak, as well as the spatial form of the differential conductance around defects. The good quantitative agreement of these theoretical results with experimental measurements provided strong and direct evidence for the proposed mechanism underlying the unconventional superconducting state in heavy fermion materials.

Extending his work to investigate the nonequilibrium properties of heavy fermion materials, Dr. Van Dyke showed that the onset of Kondo screening and the ensuing changes in the electronic structure of the material significantly alter the spatial paths of currents flowing through heavy fermion systems. The considerable experimental advances in imaging the spatial flow of currents over the last few years have therefore opened a new venue for exploring the out-of-equilibrium signatures of strong correlation effects.

In the last part of his thesis, Dr. Van Dyke investigated the nonequilibrium charge transport in a new topological state of matter, the topological insulators (TIs), which are characterized by an insulating bulk, and gapless edge or surface modes. The topological nature of these materials renders their properties robust against many forms of disorder, making them of great interest for a whole range of technological applications in quantum computation and spin-based electronics. A major hurdle in the realization of these applications has been the lack of ability to independently create and control spin and charge currents at the nanoscale. Dr. Van Dyke showed that this obstacle can be overcome and that such control can be established by breaking the time-reversal symmetry of nanoscopic TIs via magnetic defects. This symmetry breaking does not only enable one to create nearly 100% spin-polarized charge currents, but it also allows for the design of novel spin diodes. The flow of spin and charge in these diodes can be controlled at the nanoscale by changing the gate and bias voltages, which provides the missing link in the use of TIs for technological applications. These results open unprecedented opportunities to employ nanoscale TIs for applications in spintronics and quantum information.

The study of topological and strongly correlated materials will continue to fascinate physicists for years to come, and Dr. Van Dyke's thesis provides a nice introduction into these exciting fields of research.

Chicago, IL, USA Dirk Morr
October 2017

Contents

1 **Introduction** 1
   1.1 Correlations in Condensed Matter 1
   1.2 Topological Materials 5
   References 6

2 **Superconducting Gap in $CeCoIn_5$** 9
   2.1 Superconducting Gap Symmetry 9
   2.2 Basics of Scanning Tunneling and Quasiparticle Interference Spectroscopy 9
   2.3 Experimental Challenge of QPI for $CeCoIn_5$ 11
   2.4 Theoretical Model for $CeCoIn_5$ Band Structure 12
   2.5 Theory of Heavy Fermion QPI 16
   2.6 $CeCoIn_5$ QPI at Large Energies 18
   2.7 $CeCoIn_5$ QPI at Small Energies 20
   References 26

3 **Pairing Mechanism in $CeCoIn_5$** 29
   3.1 Heavy Fermion Superconductivity 29
   3.2 Extraction of the Magnetic Interaction 30
   3.3 Phase-Sensitive QPI 36
   3.4 Spin Excitations in $CeCoIn_5$ 39
      3.4.1 Magnetic Resonance Peak 40
      3.4.2 NMR Spin-Lattice Relaxation Rate 43
   References 44

4 **Real and Momentum Space Probes in $CeCoIn_5$: Defect States in Differential Conductance and Neutron Scattering Spin Resonance** ... 47
   4.1 Real-Space Study of Defects by STM 47
      4.1.1 Model 47
   4.2 Neutron Scattering in $CeCoIn_5$ 52
      4.2.1 Magnetic Anisotropy and External Magnetic Field 59
   References 62

5	**Transport in Nanoscale Kondo Lattices**	65
	5.1 Transport in a Clean System	66
	5.2 Transport with Defects	68
	5.3 Multiple Defects	70
	5.4 Hopping Within the f-Band	71
	5.5 Self-Consistency with Finite Bias	71
	References	75
6	**Charge and Spin Currents in Nanoscale Topological Insulators**	77
	6.1 Introduction	77
	6.2 Model	78
	6.3 Polarized Spin Currents	78
	6.4 Non-magnetic Defects	82
	6.5 Magnetic Defects	82
	6.5.1 Ising-Type Magnetic Defects	83
	6.5.2 Spin-Flip-Type Magnetic Defects	85
	6.6 Heisenberg Defects and Spin Diodes	86
	6.7 Interface with Ferro- and Antiferromagnets	89
	6.8 Robustness of the Spin-Polarized Currents	90
	References	96
7	**Conclusion**	97
	References	98
	Appendix A Keldysh Formalism for Transport	99
	References	102

Published Results and Contribution of Authors

Parts of this thesis have been published in the following journal articles:

[1]. M.P. Allan, F. Massee, D.K. Morr, J. Van Dyke, A.W. Rost, A.P. Mackenzie, C. Petrovic, J.C. Davis, Imaging Cooper pairing of heavy fermions in $CeCoIn_5$. Nat. Phys. **9**(8), 468–473 (2013)
[2]. J.S. Van Dyke, F. Massee, M.P. Allan, J.C.S. Davis, C. Petrovic, D.K. Morr, Direct evidence for a magnetic f-electron-mediated pairing mechanism of heavy-fermion superconductivity in $CeCoIn_5$. Proc. Natl. Acad. Sci. **111**(32), 11663–11667 (2014)
[3]. J.S. Van Dyke, J.C.S. Davis, D.K. Morr, Differential conductance and defect states in the heavy-fermion superconductor $CeCoIn_5$. Phys. Rev. B **93**(4), 041107 (2016)
[4]. Y. Song, J.V. Dyke, I.K. Lum, B.D. White, S. Jang, D. Yazici, L. Shu, A. Schneidewind, P. Čermák, Y. Qiu, M.B. Maple, D.K. Morr, P. Dai, Robust upward dispersion of the neutron spin resonance in the heavy fermion superconductor $Ce_{1-x}Yb_xCoIn_5$. Nat. Commun. **7**, 12774 (2016)
[5]. J.S. Van Dyke, D.K. Morr, Controlling the flow of spin and charge in nanoscopic topological insulators. Phys. Rev. B **93**(8), 081401 (2016)
[6]. J.S. Van Dyke, D.K. Morr, Effects of defects and dephasing on charge and spin currents in two-dimensional topological insulators. Phys. Rev. B **95**(4), 045151 (2017)

Chapter 1 is a brief introduction to the research areas considered in the thesis. Chapters 2 and 3 are based on [1] and [2]. C.P. synthesized and characterized the samples. M.P.A. and F.M. performed the STM experiments, prepared the data, and produced the published versions of the figures. J.V. and D.K.M. developed the band structure and superconducting gap models. J.V. performed the theoretical calculations discussed in the text. J.C.D. and D.K.M. supervised the projects. The manuscripts reflect the important contributions of all the authors. Chapter 4 is based on the manuscripts [3] and [4]. For [3], J.V. performed the calculations, D.K.M. supervised the project, and all the authors contributed to the writing of the manuscript. For [4], I.K.L., B.D.W., S.J., D.Y., L. Shu, and M.B.M. synthesized and characterized the samples. Y.S. performed the experiments with the assistance of A.S., P.C., and Y.Q., J.V. and D.K.M. developed the theoretical models; and J.V. performed the calculations. Y.S. prepared the data, and Y.S., J.V., and D.K.M.

produced figures. P.D. directed the project. Chapter 5 is based on unpublished work by J.V. and D.K.M., in which D.K.M. supervised the project and J.V. performed the calculations and produced the figures. Chapter 6 is based on the manuscripts [5] and [6]. J.V. and D.K.M. performed the calculations, and D.K.M. supervised the project and produced the published versions of the figures.

Chapter 1
Introduction

1.1 Correlations in Condensed Matter

The field of condensed matter physics is enormous in scope, extending from the earliest developments in crystallography to cutting-edge applications of holographic dualities (inspired by string theory) to high temperature superconductors. Despite the profusion of material systems and theoretical methods, there are a number of paradigms that serve to orient much of the work in the field. One such set of principles is Landau's Fermi liquid theory, which underlies the description of ordinary metallic systems [1, 2]. Landau surmised, as was later proven by quantum field theoretical techniques, that the low-lying excitations of a system of interacting fermions can be described in terms of renormalized "quasiparticles" with the same general behavior as the non-interacting system, but having an effective mass different from that of the original particles. This approach provides a good treatment of many simple metals, however, it can fail in systems where there are strong correlations and/or reduced dimensionalities (as in the one-dimensional Luttinger liquid, for example [3]).

Much of the present thesis concerns the description of strongly correlated systems. This class of materials, which is of course smaller than condensed matter as a whole, is still extremely broad. Among correlated systems, this work deals exclusively with the particular subclass of heavy fermion systems, and a large part is devoted to the specific material $CeCoIn_5$. Before entering into details, it may be helpful to briefly discuss strongly correlated electrons in general, following Ref. [4]. One characteristic feature of strongly correlated electron systems is the presence of low energy scales that do not appear in non-interacting or weakly correlated systems (where the energy scale is set by the Fermi energy). It is important to note that, in different systems, the cause of these scales may be quite different. Thus, grouping together systems on this basis is somewhat like classifying diseases according to their symptoms. This can be a useful undertaking, but it does not necessarily lead to a cure. A concrete example of generation of a new low energy scale is given by the Kondo effect [4, 5].

The Kondo problem has a rich history, beginning with the observation of an anomalous minimum in the temperature dependence of the electrical resistance of some metals at low temperatures [6]. It was recognized early on that this minimum could be due to the presence of residual impurities in the host, which was confirmed by studies demonstrating the change in the location of the minimum under the controlled addition of defects [7]. The theoretical explanation for the minimum was pioneered by Jun Kondo [8], who calculated the scattering of conduction electrons by a magnetic impurity to third order in perturbation theory, thereby showing a log divergence of the scattering rate with the inverse temperature. This explained the appearance of the resistance minimum (when the phonon contribution, which decreases with temperature, is overcome by the magnetic scattering term), but left open the question of what happens at still lower temperatures where the perturbation expansion breaks down. This became known as the Kondo problem. Many groups contributed to the understanding of the problem, but perhaps the most crucial physical idea came from Anderson's scaling theory [9, 10], which suggested that the increased coupling between the conduction electrons in the metal and the localized magnetic moment eventually leads to the formation of a singlet bound state between the two. The temperature at which this crossover takes place is known as the Kondo temperature, T_K. Anderson's ideas were confirmed by Wilson, using his numerical renormalization group approach [11]. Later studies of the problem included an effective local Fermi liquid approach [12], conformal field theories [13], large-N expansions [14–16], and even exact solutions via the Bethe ansatz [17, 18].

In the course of this development, several different theoretical models were proposed and studied to shed light on the experimental results in Kondo impurity systems. The Hamiltonian studied by Kondo (now often called the Kondo model) represents the antiferromagnetic interaction between the spins of the local moment and the conduction electrons:

$$H = J \sum_{\mathbf{k}} \mathbf{s_k} \cdot \mathbf{S}_{\text{imp}} \tag{1.1}$$

Here $\mathbf{s_k}$ and $\mathbf{S}_{\text{imp}}$ represent the spin operators of the conduction electrons and the local moment, respectively. This model, like many used in heavy fermion physics, has a deceptive simplicity to it. In fact, the use of spin operators prevents the straightforward application of quantum field theory techniques, since the commutation relations imposed on spins do not admit a Wick theorem [5]. Although alternative perturbation theories can be developed [19], the standard procedure is to re-write the spin operators in terms of bosonic or fermionic operators and a constraint. The choice of operators is typically a matter of convenience for whatever problem is at hand. In magnetic phases, bosonic representations have been found useful, whereas studies of Fermi liquid states have tended to use fermionic ones [20]. In the Kondo problem, the constraint has the physical interpretation that the magnetic moment arises from a localized (usually f or d) electron at the impurity site; that is, charge fluctuations are neglected.

1.1 Correlations in Condensed Matter

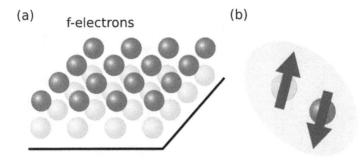

Fig. 1.1 Schematic drawing of (**a**) layered heavy fermion system (**b**) Kondo screening of f-electron spins by conduction electrons. The yellow and red spheres represent the f-electrons and conduction electrons, respectively

Another widely used model for the single impurity problem is the eponymous Anderson model, which is more general in that charge fluctuations are permitted. The Hamiltonian in this case can be written as

$$H = \sum_{\mathbf{k},\sigma} \varepsilon_{\mathbf{k}}^c c_{\mathbf{k},\sigma}^\dagger c_{\mathbf{k},\sigma} + E_0 \sum_\sigma n_\sigma^f + U n_\uparrow^f n_\downarrow^f + \sum_{\mathbf{k},\sigma} V_{\mathbf{k}} f_\sigma^\dagger c_{\mathbf{k},\sigma} + H.c. \quad (1.2)$$

where $c_{\mathbf{k},\sigma}^\dagger$ ($c_{\mathbf{k},\sigma}$) creates (annihilates) a conduction electron with momentum $\mathbf{k}$ and spin σ, $f_\sigma^\dagger$ (f_σ) creates (annihilates) a localized f-electron with spin σ, and the operator $n_\sigma^f = f_\sigma^\dagger f_\sigma$ gives the number of f-electrons at the impurity site with a given spin. Furthermore, the dispersion of the conduction electrons is given by $\varepsilon_{\mathbf{k}}^c$, E_0 is an on-site energy for the f-electron on the impurity, U describes the Coulomb repulsion between electrons at the impurity (Hubbard potential), and $V_{\mathbf{k}}$ is the hybridization between the c- and f-electrons. This model is more complex than the Kondo model, but a definite relation exists between the two, as follows. Taking the Coulomb repulsion $U \to \infty$, double occupation of the f-electron site is forbidden and the original Hilbert space is projected down to the subspace of states in which the site is either unoccupied or singly occupied. Requiring further that the occupation $n_\uparrow^f + n_\downarrow^f = 1$, one can perform a Schrieffer-Wolff transformation to recover the Kondo model [21].

The generalization of the Kondo effect to a lattice of local moments yields one of the classic examples of strongly correlated systems, namely, a heavy fermion system. These materials also provide an instance where Fermi liquid behavior can survive in the presence of strong correlations. At high temperatures, a heavy fermion system can be modeled as a lattice of localized magnetic f-electrons (spins) interacting with a band of conduction electrons, as shown schematically in Fig. 1.1a. As the temperature is lowered, there is a crossover to a new state where the conduction electrons screen the local moments (Fig. 1.1b), producing a Fermi liquid of residual non-magnetic quasiparticles. These quasiparticles can have very large

Fig. 1.2 Schematic drawing of the crystal structure of CeCoIn$_5$ [28]

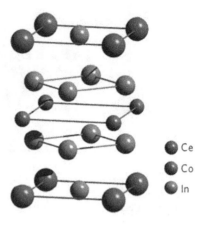

effective masses, up to more than a thousand times the bare mass of the electron (hence the name "heavy fermion"). The emergence of the low temperature Fermi liquid is a surprising and beautiful example of universality in condensed matter systems, wherein complicated microscopic physics gives rise to relatively simple behavior at low energies. In general, heavy fermion materials have complex phase diagrams which can also include magnetism and superconductivity in addition to Fermi liquid physics. In some cases there are even regions—known as non-Fermi liquid phases or strange metals—that are not well-described within any of the traditional paradigms of condensed matter physics, but are instead believed to be associated with a quantum phase transition at zero temperature [22, 23].

The first part of this thesis focuses on the unconventional superconducting state of CeCoIn$_5$, one of the most perplexing phenomena in heavy fermion materials. Superconductivity was found in this compound at 2.3 K, giving it the highest transition temperature of the Ce-based materials. The discovery marked the beginning of a concerted effort to understand its normal and superconducting states [24]. CeCoIn$_5$ has a tetragonal crystal structure [24], as shown schematically in Fig. 1.2. There exist two compounds related to CeCoIn$_5$ which are isostructural to it: CeRhIn$_5$ [25] and CeIrIn$_5$ [26]. The various similarities and differences between the three materials are helpful for understanding each in its own light as well. The "Ce-115 compounds" are also believed to be related to the cuprate and iron pnictide superconductors, due to their similar quasi-2D structures, unconventional superconductivity, and proximity to antiferromagnetic states [27]. A better understanding of superconductivity in heavy fermions like CeCoIn$_5$ will also likely shed light on the complex physical properties of the high temperature superconductors, which have resisted a complete description for nearly 30 years.

The significant advances discussed in the present work were made possible by cutting-edge scanning tunneling microscopy experiments (STM) on CeCoIn$_5$, performed by the Davis group at Cornell University [28, 29]. These experiments allowed for the extraction of the low-energy electronic structure by the method of quasiparticle interference spectroscopy, which is crucial for developing a quantitative understanding of superconductivity in the material (Chap. 2). Together with the detailed form of the magnetic interaction between f-electrons, also extracted from the experiments, this information led to a series of predictions about the superconducting state, including the determination of the gap symmetry and critical temperature (Chap. 3). This was achieved primarily on the basis of experimental input relevant for the normal, as opposed to the superconducting, state of the material. Furthermore, the model developed in this work has found use in the study of the local response in STM experiments to the presence of defects, as well as in the analysis of recent neutron scattering experiments (Chap. 4). The results discussed so far all deal with equilibrium properties of heavy fermion materials. Nonequilibrium experiments also pose exciting challenges and opportunities for advancing the theoretical understanding of correlated systems. To this end, a simplified model of a nanoscale heavy fermion system is studied to determine the currents that flow through the sample in the presence of an applied voltage. The effect of defects and correlations on the current patterns are examined, as well as the role of the non-zero bias on the correlations themselves (Chap. 5).

1.2 Topological Materials

Continuing with the study of current flow, the final chapter turns to the behavior of nanoscale topological insulators (TIs). In many cases these materials can be understood in terms of non-interacting physics, but they are nonetheless currently of great interest due to the existence of certain topological invariants that can be used to characterize distinct states of the system [30, 31]. The values of these invariants are quantized, and cannot change between two regions of space without closing the insulating gap at the Fermi level. Thus, if the material is in a state corresponding to a nontrivial topological invariant, the fact that the vacuum is an insulator (with trivial invariant) implies that the system possesses conducting surface or edge states. Furthermore, the edge states are spin-momentum locked, in the sense that electrons of a given spin are forced to travel in a particular direction around the edge, while those of the opposite spin propagate in the opposite direction (Fig. 1.3). This is true not only in a clean system, but even in the presence of nonmagnetic defects. In this

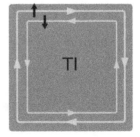

Fig. 1.3 Schematic drawing of a topological insulator illustrating spin-momentum locking. Spin-↑ electrons travel clockwise around the edge, whereas spin-↓ electrons travel counterclockwise

case, time-reversal symmetry protects against backscattering—roughly speaking, for any given backscattering trajectory, there is another related by time-reversal symmetry which interferes destructively with the first [31].

However, by introducing magnetic impurities on the edge of a 2D TI, one explicitly breaks this symmetry and backscattering may occur. We demonstrate how one can generate highly spin-polarized currents using magnetic defects appropriately placed on the surface of a 2D TI. To bolster support for this claim, we show that the results are robust against various perturbations of the model and that similar effects can be achieved by interfacing at TI with a disordered ferromagnet or an antiferromagnet. The generation of spin-polarized currents is an important goal for the development of next-generation technology in the fields of spintronics and quantum computing. Hence, we anticipate that the proposal outlined here will find use in future applications.

References

1. L.D. Landau, The theory of a Fermi liquid. Sov. Phys. J. Exp. Theor. Phys. **3**, 920 (1957)
2. A.A. Abrikosov, L.P. Gorkov, I.E. Dzyaloshinskii, *Methods of Quantum Field Theory in Statistical Physics*, Revised English edn. (Dover Publications, New York, 1975)
3. T. Giamarchi, *Quantum Physics in One Dimension*, 1st edn. (Clarendon Press, New York, 2004)
4. P. Fulde, P. Thalmeier, G. Zwicknagl, Strongly correlated electrons, in *Solid State Physics*, vol. 60 (Elsevier, San Diego, 2006)
5. A.C. Hewson, *The Kondo Problem to Heavy Fermions* (Cambridge University Press, Cambridge, 1997)
6. W.J. de Haas, J. de Boer, G.J. van dën Berg, The electrical resistance of gold, copper and lead at low temperatures. Physica **1**(7), 1115–1124 (1934)
7. D.K.C. MacDonald, W.B. Pearson, I.M. Templeton, Thermo-electricity at low temperatures. IX. The transition metals as solute and solvent. Proc. R. Soc. Lond. A Math. Phys. Eng. Sci. **266**(1325), 161–184 (1962)
8. J. Kondo, Resistance minimum in dilute magnetic alloys. Prog. Theor. Phys. **32**(1), 37–49 (1964)
9. P.W. Anderson, G. Yuval, D.R. Hamann, Exact results in the Kondo problem. II. Scaling theory, qualitatively correct solution, and some new results on one-dimensional classical statistical models. Phys. Rev. B **1**(11), 4464–4473 (1970)
10. P.W. Anderson, A poor man's derivation of scaling laws for the Kondo problem. J. Phys. C Solid State Phys. **3**(12), 2436 (1970)
11. K.G. Wilson, The renormalization group: critical phenomena and the Kondo problem. Rev. Mod. Phys. **47**(4), 773–840 (1975)
12. P. Nozières, A "fermi-liquid" description of the Kondo problem at low temperatures. J. Low Temp. Phys. **17**(1–2), 31–42 (1974)
13. I. Affleck, A.W.W. Ludwig, Exact conformal-field-theory results on the multichannel Kondo effect: single-fermion Green's function, self-energy, and resistivity. Phys. Rev. B **48**(10), 7297–7321 (1993)
14. P. Coleman, $\frac{1}{N}$ expansion for the Kondo lattice. Phys. Rev. B **28**(9), 5255–5262 (1983)
15. N. Read, D.M. Newns, On the solution of the Coqblin-Schreiffer Hamiltonian by the large-N expansion technique. J. Phys. C Solid State Phys. **16**(17), 3273 (1983)

References

16. D.M. Newns, N. Read, Mean-field theory of intermediate valence/heavy fermion systems. Adv. Phys. **36**(6), 799–849 (1987)
17. N. Andrei, Diagonalization of the Kondo Hamiltonian. Phys. Rev. Lett. **45**(5), 379–382 (1980)
18. P.B. Wiegmann, Exact solution of s-d exchange model at T=0. Sov. Phys. J. Exp. Theor. Phys. Lett. **31**(7), 364 (1980)
19. H. Keiter, J.C. Kimball, Diagrammatic perturbation technique for the Anderson Hamiltonian, and relation to the sd exchange Hamiltonian. Int. J. Magn. **1**, 233 (1971)
20. A. Ramires, P. Coleman, Supersymmetric approach to heavy fermion systems. Phys. Rev. B **93**(3), 035120 (2016)
21. J.R. Schrieffer, P.A. Wolff, Relation between the Anderson and Kondo Hamiltonians. Phys. Rev. **149**(2), 491–492 (1966)
22. V.A. Sidorov, M. Nicklas, P.G. Pagliuso, J.L. Sarrao, Y. Bang, A.V. Balatsky, J.D. Thompson, Superconductivity and quantum criticality in $CeCoIn_5$. Phys. Rev. Lett. **89**(15), 157004 (2002)
23. S. Sachdev, *Quantum Phase Transitions*, 2nd edn. (Cambridge University Press, Cambridge, 2011)
24. C. Petrovic, P.G. Pagliuso, M.F. Hundley, R. Movshovich, J.L. Sarrao, J.D. Thompson, Z. Fisk, P. Monthoux, Heavy-fermion superconductivity in $CeCoIn_5$ at 2.3 K. J. Phys. Condens. Matter **13**(17), L337 (2001)
25. H. Hegger, C. Petrovic, E.G. Moshopoulou, M.F. Hundley, J.L. Sarrao, Z. Fisk, J.D. Thompson, Pressure-induced superconductivity in quasi-2D $CeRhIn_5$. Phys. Rev. Lett. **84**(21), 4986–4989 (2000)
26. C. Petrovic, R. Movshovich, M. Jaime, P.G. Pagliuso, M.F. Hundley, J.L. Sarrao, Z. Fisk, J.D. Thompson, A new heavy-fermion superconductor $CeIrIn_5$: a relative of the cuprates? Europhys. Lett. **53**(3), 354–359 (2001)
27. D.J. Scalapino, A common thread: the pairing interaction for unconventional superconductors. Rev. Mod. Phys. **84**(4), 1383–1417 (2012)
28. M.P. Allan, F. Massee, D.K. Morr, J. Van Dyke, A.W. Rost, A.P. Mackenzie, C. Petrovic, J.C. Davis, Imaging Cooper pairing of heavy fermions in $CeCoIn_5$. Nat. Phys. **9**(8), 468–473 (2013)
29. J.S. Van Dyke, F. Massee, M.P. Allan, J.C.S. Davis, C. Petrovic, D.K. Morr, Direct evidence for a magnetic f-electron-mediated pairing mechanism of heavy-fermion superconductivity in $CeCoIn_5$. Proc. Natl. Acad. Sci. **111**(32), 11663–11667 (2014)
30. M.Z. Hasan, C.L. Kane, *Colloquium*: topological insulators. Rev. Mod. Phys. **82**(4), 3045–3067 (2010)
31. X.-L. Qi, S.-C. Zhang, Topological insulators and superconductors. Rev. Mod. Phys. **83**(4), 1057–1110 (2011)

Chapter 2
Superconducting Gap in CeCoIn$_5$

2.1 Superconducting Gap Symmetry

One of the central questions that can be asked about any bulk superconductor is the symmetry of its superconducting gap, $\Delta(\mathbf{k})$. This gap in the excitation spectrum of the superconductor is a consequence of the finite energy required to break apart a Cooper pair [1]. Following the standard BCS theory, the gap is a function of the momenta of the electrons forming the Cooper pair. While the elemental superconductors and those composed of simple alloys invariably possess s-wave symmetry ($\Delta(\mathbf{k}) = const.$), more complicated systems such as the cuprates, iron pnictides, and heavy fermions can possess other symmetries of their gap functions [2]. In the case of CeCoIn$_5$, numerous experimental studies have been undertaken to try to determine the symmetry of the gap. Early measurements of the angular dependence of the thermal conductivity showed a fourfold symmetry indicative of $d_{x^2-y^2}$ pairing [3]. Some time later, the magnetic field angle dependence of the specific heat was also found to have a fourfold symmetry, but one that was suggestive of d_{xy} instead [4]. Thus, although the superconductivity was likely to be unconventional (i.e., allowing for a change in the phase of the gap as a function of momentum), it was unclear exactly what was the symmetry of the gap. Furthermore, these thermodynamic studies of the gap do not constitute direct observations of the gap, but rather rely on theoretical interpretation, which is uncertain.

2.2 Basics of Scanning Tunneling and Quasiparticle Interference Spectroscopy

A major advance came with the advent of scanning tunneling spectroscopy (STS) experiments on CeCoIn$_5$ [5, 6]. In these experiments, a scanning tunneling microscope (STM) is used to probe the electronic structure at the surface of the material.

Fig. 2.1 Schematic drawing of an STM experiment, illustrating how electrons tunnel between the tip and sample surface

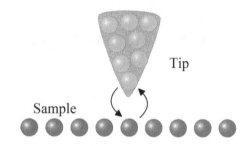

While the STM was originally developed with the goal of imaging surfaces with atomic resolution, thereby providing a topographic map of step edges, adsorbates, and other surface phenomena, the STS mode of operation has become a powerful method of directly studying the underlying electronic structure of materials as well [7–9]. Briefly, in an STS experiment the STM tip is fixed above a given atomic site and the voltage bias between the tip and sample is varied. Electrons are able to tunnel between the tip and sample through the insulating barrier of the vacuum (Fig. 2.1). One records the differential conductance (dI/dV) as a function of bias V, the former quantity being proportional to the local density of states (for a single band in the weak tunneling limit) [10].

$$\frac{dI(\mathbf{r}, E)}{dV} = \frac{2\pi e}{\hbar} N_t t^2 N_s(\mathbf{r}, E) \tag{2.1}$$

Here N_t is the density of states of the STM tip, t is the hopping integral between the tip and sample, and $N_s(\mathbf{r}, E)$ is the local density of states of the material at position $\mathbf{r}$. The measurement is repeated at every site in a two-dimensional field of view on the sample surface. The tunneling data as a function of energy reveals important information about the electronic structure of the material. One may detect such features as van Hove singularities arising from the flatness of a given band, or superconducting [11] or hybridization [12] gaps around the Fermi level.

The usefulness of the STS technique was extended further with the introduction of quasiparticle interference (QPI) spectroscopy [13–15]. It is well-known that the placement of a charged impurity in a homogeneous electron gas leads to oscillations of the electron density, as the gas attempts to screen the perturbing charge [16]. These oscillations possess a characteristic wavevector of $2k_F$, due to the scattering of the electrons across the Fermi surface (the Fermi sphere in the non-interacting case). These effects has been observed in solids as well and are known as Friedel oscillations [17], wherein the conduction electrons scatter off defects in the material, resulting in ripple-like spatial patterns in the charge density (and correspondingly in the local density of states) around the defect. The oscillations are often readily observed in STM surface maps at fixed bias, and by Fourier transforming the 2D real space image, one extracts the principal wavevectors $\mathbf{q}$ that occur in them. Since the energy of the probed quasiparticle states is determined by the bias through $E = eV$, one obtains $\mathbf{q}(E)$, the transferred momentum as a function of quasiparticle energy.

Under the further approximation of a spherical Fermi surface, $\mathbf{q} = 2\mathbf{k}$, the foregoing relation can be inverted to give an experimental determination of the electronic band structure, $E(\mathbf{k})$ [13, 14].

One might inquire as to the validity of the band structure extracted using the above procedure, since the STM is sensitive only to states near the surface of the material, and furthermore only determines the scattering wavevector in the plane (whereas in the bulk the bands are generically dispersing in three dimensions). In reply, it should be kept in mind that the systems to which this technique is applied, including $CeCoIn_5$, are quasi-2D in nature, with the important electronic bands lying within the plane. Thus, there is good reason to believe that the electronic structure at the surface is also indicative of the bulk physics, and so the QPI method can provide insight into the general behavior of such materials.

2.3 Experimental Challenge of QPI for $CeCoIn_5$

We now present the experimental results of Allan et al. [5] and show how they can be successfully understood using the periodic Anderson model, one of the theoretical cornerstones for describing the complex physics of heavy fermion materials. Figure 2.2 shows three examples of QPI data obtained from a measurement on a sample of $CeCoIn_5$. As always found for QPI experiments, there is a strong background of intensity near $\mathbf{q} = (0, 0)$ arising from large-scale surface modulations. The relevant points in the data for extracting the quasiparticle dispersions are the regions of high intensity located at larger wavevectors; these are indicated in the figure by the numbered circles. As the bias is varied, it is possible to reliably and reproducibly track the movement of these spots in the $\mathbf{q}$-space. In particular, one may focus on one-dimensional cuts along two of the high symmetry directions in the Brillouin zone, $(0, 0) \to (0, 2\pi/a_0)$ and $(0, 0) \to (2\pi/a_0, 2\pi/a_0)$. The goal of the theorist is to employ the evolution of the QPI scattering maxima as a function of energy to extract the material's underlying electronic structure. The experimental QPI cuts recorded at a temperature of 250 mK are shown in Fig. 2.3.

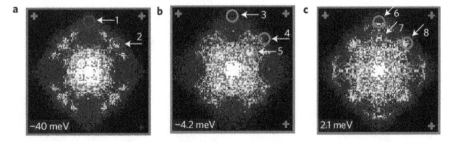

Fig. 2.2 Sample experimental data obtained from QPI analysis of STS measurements on $CeCoIn_5$ [5]. Numbered circles indicate important scattering wavevectors (**a**) −40 meV, (**b**) −4.2 meV, (**c**) 2.1 meV

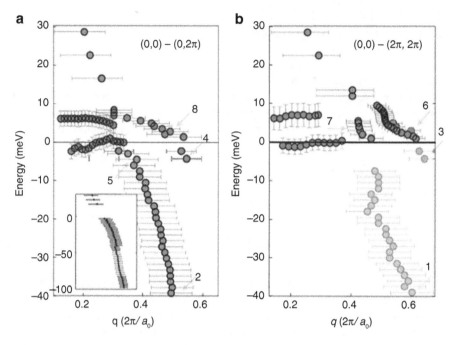

Fig. 2.3 One-dimensional cuts along the directions (a) $(0, 0) \to (0, 2\pi/a_0)$ and (b) $(0, 0) \to (2\pi/a_0, 2\pi/a_0)$ of QPI analysis on CoIn$_5$ [5]. Numbered arrows indicate important scattering wavevectors shown in Fig. 2.2

One immediately notices that it is possible to distinguish between scattering due to the "light" and "heavy" parts of the bands. The scattering from the light bands is highly dispersive, as indicated by the gray circles, and arises from the contribution of the delocalized conduction electrons in the system ("light" refers to the fact that the effective mass of the quasiparticles does not differ greatly from the bare mass of the electron). On the other hand, the flat regions in the scattering plot, shown with blue circles, come from the heavy portions of the hybridized quasiparticles bands. In this region, the quasiparticles have very large effective masses due to the strong correlations between the conduction electrons and localized f-electrons of the Ce atoms.

2.4 Theoretical Model for CeCoIn$_5$ Band Structure

To try to reproduce the experimental findings presented above one would like to develop a quantitative theoretical description of the electronic structure. The class of models appropriate for heavy fermion systems like CeCoIn$_5$ has a long history dating back to the discovery of the first materials in the 1970s [18], up through their continued theoretical elucidation today (see [5] of Chap. 1). These models are essentially the extension to the lattice of the models designed to describe individual magnetic impurities in host metals, the Kondo problem discussed in the previous chapter.

2.4 Theoretical Model for CeCoIn$_5$ Band Structure

In heavy fermion systems with dense lattices of local moments, interactions between neighboring moments invalidate the simple single-impurity models discussed previously. In this case, the natural generalizations of models (1.1) and (1.2) are

$$H = J \sum_{\mathbf{r}} \mathbf{S}_{\mathbf{r}} \cdot \mathbf{S}_{\text{imp}}(\mathbf{r}) \tag{2.2}$$

and

$$H = \sum_{\mathbf{k},\sigma} \varepsilon^c_{\mathbf{k}} c^\dagger_{\mathbf{k},\sigma} c_{\mathbf{k},\sigma} + E_0 \sum_{\mathbf{r}} n^f_{\mathbf{r}} + \sum_{\mathbf{r},\mathbf{r}'} I_{\mathbf{r},\mathbf{r}'} \mathbf{S}_{\mathbf{r}} \cdot \mathbf{S}_{\mathbf{r}'} + \sum_{\mathbf{r},\mathbf{r}',\sigma} V_{\mathbf{r},\mathbf{r}'} f^\dagger_{\mathbf{r},\sigma} c_{\mathbf{r}',\sigma} + \text{H.c.} \tag{2.3}$$

We have further added a Heisenberg-like term $H_H = \sum_{\mathbf{r},\mathbf{r}'} I_{\mathbf{r},\mathbf{r}'} \mathbf{S}_{\mathbf{r}} \cdot \mathbf{S}_{\mathbf{r}'}$ to allow for interactions between neighboring magnetic moments. In the Kondo limit $n^f = 1$, the Anderson lattice model can be mapped onto the Kondo lattice model. Anticipating this possibility that n^f may differ from one for CeCoIn$_5$, we choose to work with the periodic Anderson model in the following.

We now turn to the approximate solution of this model (in the infinite U limit). The exclusion of double occupancy on the f site can be conveniently described in the slave boson approach [19–22]. One introduces a set of new bosonic operators $b_{\mathbf{r}}, b^\dagger_{\mathbf{r}}$ to label unoccupied sites. The hybridization terms, which transfer electrons into or out of the f orbital, are changed via $V_{\mathbf{r},\mathbf{r}'} f^\dagger_{\mathbf{r},\sigma} c_{\mathbf{r}',\sigma} \to V_{\mathbf{r},\mathbf{r}'} f^\dagger_{\mathbf{r},\sigma} b_{\mathbf{r}} c_{\mathbf{r}',\sigma}$ so that the Hamiltonian in the slave boson method becomes

$$H = \sum_{\mathbf{k},\sigma} \varepsilon^c_{\mathbf{k}} c^\dagger_{\mathbf{k},\sigma} c_{\mathbf{k},\sigma} + E_0 \sum_{\mathbf{r}} n^f_{\mathbf{r}} + \sum_{\mathbf{r},\mathbf{r}',\sigma} V_{\mathbf{r},\mathbf{r}'} f^\dagger_{\mathbf{r},\sigma} b_{\mathbf{r}} c_{\mathbf{r}',\sigma} + \text{H.c.} + \sum_{\mathbf{r},\mathbf{r}'} I_{\mathbf{r},\mathbf{r}'} \mathbf{S}_{\mathbf{r}} \cdot \mathbf{S}_{\mathbf{r}'} \tag{2.4}$$

The constraint $n^f \leq 1$ can now be represented as $b^\dagger_{\mathbf{r}} b_{\mathbf{r}} + \sum_{\sigma} f^\dagger_{\mathbf{r}} f_{\mathbf{r}} = 1$, which is to be enforced at each site $\mathbf{r}$. An explicit form of the constraint and the decoupling of the Heisenberg interaction term H_H is conveniently done in a path integral approach. First, the spin operators of the conduction and f electrons are replaced with Abrikosov pseudofermions, $\mathbf{S}_{\mathbf{r}} = \frac{1}{2} \Psi^\dagger_{\mathbf{r}} \boldsymbol{\sigma} \Psi_{\mathbf{r}}$, with spinor $\Psi^\dagger_{\mathbf{r}} = (f^\dagger_{\mathbf{r}\uparrow}, f^\dagger_{\mathbf{r}\downarrow})$, and $\boldsymbol{\sigma} = (\sigma_1, \sigma_2, \sigma_3)$ is a vector of the Pauli matrices. Next, we decouple the interaction term between f-electrons using the standard Hubbard-Stratonovich method, introducing a new field $t_f(\mathbf{r}, \mathbf{r}', \tau)$. The static approximation substitutes for this field its expectation value $t_f(\mathbf{r}, \mathbf{r}')$, and similarly for the slave boson operators $b_{\mathbf{r}} \to r_0(\mathbf{r})$. Furthermore, the f occupation constraint is enforced by a Lagrange multiplier $\lambda = \varepsilon_f - E_0$, also taken in the static approximation. By minimizing the action with respect to ε_f and $s(\mathbf{r}, \mathbf{r}') = V_{\mathbf{r},\mathbf{r}'} r_0(\mathbf{r})$, one obtains a set of self-consistency equations

$$s(\mathbf{r}, \mathbf{r}') = \frac{J_{\mathbf{r},\mathbf{r}'}}{\pi} \int_{-\infty}^{\infty} d\omega n_F(\omega) \text{Im} G_{fc}(\mathbf{r}, \mathbf{r}', \omega) \tag{2.5}$$

$$t_f(\mathbf{r}, \mathbf{r}') = -\frac{I_{\mathbf{r},\mathbf{r}'}}{\pi} \int_{-\infty}^{\infty} d\omega n_F(\omega) \text{Im} G_{ff}(\mathbf{r}, \mathbf{r}', \omega) \tag{2.6}$$

$$n_f(\mathbf{r}) = -\int_{-\infty}^{\infty} \frac{d\omega}{\pi} n_F(\omega) \text{Im} G_{ff}(\mathbf{r}, \mathbf{r}, \omega) \tag{2.7}$$

with f occupation $n_f = 1 - r_0^2$ and $J_{\mathbf{r},\mathbf{r}'} = V_{\mathbf{r},\mathbf{r}'}/(\varepsilon_f - E_0) > 0$. The effective hoppings $s(\mathbf{r}, \mathbf{r}')$ and $t_f(\mathbf{r}, \mathbf{r}')$ encode the correlations between conduction and f-electrons, and those among f-electrons, respectively. Finally, the Green's functions on the right-hand sides of Eqs. (2.5)–(2.7) are

$$G_{cc}(\mathbf{k}, \sigma, \omega) = \frac{w_\mathbf{k}^2}{\omega - E_\mathbf{k}^\alpha + i\Gamma_\alpha} + \frac{x_\mathbf{k}^2}{\omega - E_\mathbf{k}^\beta + i\Gamma_\beta} \tag{2.8}$$

$$G_{ff}(\mathbf{k}, \sigma, \omega) = \frac{x_\mathbf{k}^2}{\omega - E_\mathbf{k}^\alpha + i\Gamma_\alpha} + \frac{w_\mathbf{k}^2}{\omega - E_\mathbf{k}^\beta + i\Gamma_\beta} \tag{2.9}$$

$$G_{cf}(\mathbf{k}, \sigma, \omega) = w_\mathbf{k} x_\mathbf{k} \left[\frac{1}{\omega - E_\mathbf{k}^\alpha + i\Gamma_\alpha} - \frac{1}{\omega - E_\mathbf{k}^\beta + i\Gamma_\beta} \right] \tag{2.10}$$

where the further assumption was made that $s(\mathbf{r}, \mathbf{r}') = s(\mathbf{r} - \mathbf{r}')$ and $t_f(\mathbf{r}, \mathbf{r}') = t_f(\mathbf{r} - \mathbf{r}')$, which may then be Fourier transformed into momentum space along with the Green's functions. Furthermore, $\Gamma_{\alpha,\beta}$ is the inverse lifetime of the heavy quasiparticles labeled by α, β and the coherence factors $w_\mathbf{k}$ and $x_\mathbf{k}$ are

$$w_\mathbf{k}^2 = \frac{1}{2} \left[1 + \frac{\left(\frac{\varepsilon_\mathbf{k}^c - \varepsilon_\mathbf{k}^f}{2}\right)^2}{\sqrt{\left(\frac{\varepsilon_\mathbf{k}^c - \varepsilon_\mathbf{k}^f}{2}\right)^2 + s_\mathbf{k}^2}} \right] \tag{2.11}$$

$$x_\mathbf{k}^2 = \frac{1}{2} \left[1 - \frac{\left(\frac{\varepsilon_\mathbf{k}^c - \varepsilon_\mathbf{k}^f}{2}\right)^2}{\sqrt{\left(\frac{\varepsilon_\mathbf{k}^c - \varepsilon_\mathbf{k}^f}{2}\right)^2 + s_\mathbf{k}^2}} \right] \tag{2.12}$$

$$w_\mathbf{k} x_\mathbf{k} = \frac{s_\mathbf{k}^2}{2\sqrt{\left(\frac{\varepsilon_\mathbf{k}^c - \varepsilon_\mathbf{k}^f}{2}\right)^2 + s_\mathbf{k}^2}} \tag{2.13}$$

2.4 Theoretical Model for CeCoIn5 Band Structure

the energies of the quasiparticle states are finally given by

$$E_{\mathbf{k}}^{\alpha,\beta} = \frac{\varepsilon_{\mathbf{k}}^c + \varepsilon_{\mathbf{k}}^f}{2} \pm \sqrt{\left(\frac{\varepsilon_{\mathbf{k}}^c - \varepsilon_{\mathbf{k}}^f}{2}\right)^2 + s_{\mathbf{k}}^2} \quad (2.14)$$

These equations give the band structure for the heavy quasiparticles in the hybridized Kondo lattice. Note that the f-electrons acquire a dispersion $\varepsilon_{\mathbf{k}}^f$ due to the hopping induced by the Heisenberg term in the Hamiltonian (2.4).

An alternative and instructive way of viewing these results is in the Hamiltonian language, wherein the static approximation amounts to the mean-field Hamiltonian

$$H_{MF} = \sum_{\mathbf{k},\sigma} \varepsilon_{\mathbf{k}}^c c_{\mathbf{k},\sigma}^\dagger c_{\mathbf{k},\sigma} + \sum_{\mathbf{k},\sigma} \varepsilon_{\mathbf{k}}^f f_{\mathbf{k},\sigma}^\dagger f_{\mathbf{k},\sigma} + \sum_{\mathbf{k},\sigma} s_{\mathbf{k}} f_{\mathbf{k},\sigma}^\dagger c_{\mathbf{k},\sigma} + H.c. \quad (2.15)$$

This non-interacting Hamiltonian is diagonalized by the following canonical transformation

$$c_{\mathbf{k},\sigma}^\dagger = w_{\mathbf{k}} \alpha_{\mathbf{k},\sigma}^\dagger + x_{\mathbf{k}} \beta_{\mathbf{k},\sigma}^\dagger \quad (2.16)$$

$$f_{\mathbf{k},\sigma}^\dagger = -x_{\mathbf{k}} \alpha_{\mathbf{k},\sigma}^\dagger + w_{\mathbf{k}} \beta_{\mathbf{k},\sigma}^\dagger \quad (2.17)$$

where $w_{\mathbf{k}}, x_{\mathbf{k}}$ come from Eqs. (2.11) and (2.12), and the final diagonalized form of the Hamiltonian is

$$H_K^{MF} = \sum_{\mathbf{k},\sigma} \left(E_{\mathbf{k}}^\alpha \alpha_{\mathbf{k},\sigma}^\dagger \alpha_{\mathbf{k},\sigma} + E_{\mathbf{k}}^\beta \beta_{\mathbf{k},\sigma}^\dagger \beta_{\mathbf{k},\sigma} \right) \quad (2.18)$$

At this point we have a model for the excitations of the low temperature, hybridized heavy Fermi liquid state, but without superconductivity. To study the superconducting state, one may proceed in two different ways. The more theoretical approach, developed in Chap. 3 is to introduce a superconducting pairing interaction between the quasiparticles which could be used along with the band structure to determine the properties of the superconducting state in a weak-coupling BCS approach. However, this requires a definite proposal for the microscopic pairing mechanism, for which there are multiple possibilities [23–26]. It may be that several different mechanisms are capable of accounting for the observed QPI, so in the present chapter we restrict ourselves to the issues that can be addressed independently of the choice of mechanism.

No matter what the fundamental nature of the pairing mechanism, superconductivity can be incorporated in a model Hamiltonian at the mean-field (BCS) level by the addition of pairing terms:

$$H_{SC}^{MF} = -{\sum_p}' (\Delta_{\mathbf{k}}^\alpha \alpha_{\mathbf{k},\downarrow} \alpha_{-\mathbf{k},\uparrow} + \Delta_{\mathbf{k}}^\beta \beta_{\mathbf{k},\downarrow} \beta_{-\mathbf{k},\uparrow} + H.c.) \quad (2.19)$$

where the prime on the summation indicates a restriction to states within the Debye energy of the Fermi energy

$$|E_k^{\alpha,\beta}| \leq \omega_D \qquad (2.20)$$

Then the total mean-field Hamiltonian in the superconducting state is

$$H_{tot}^{MF} = H_K^{MF} + H_{SC}^{MF} \qquad (2.21)$$

The Hamiltonian is off-diagonal on account of the pairing terms, but can be diagonalized with the canonical (Bogoliubov) transformations

$$\alpha_{\mathbf{k},\uparrow} = u_{\mathbf{k}}^\alpha a_{\mathbf{k}} + v_{\mathbf{k}}^\alpha b_{\mathbf{k}}^\dagger \qquad (2.22)$$

$$\alpha_{-\mathbf{k},\downarrow} = v_{\mathbf{k}}^\alpha a_{\mathbf{k}}^\dagger - u_{\mathbf{k}}^\alpha b_{\mathbf{k}} \qquad (2.23)$$

for the α-band, while for the β-band one has

$$\beta_{\mathbf{k},\uparrow} = u_{\mathbf{k}}^\beta d_{\mathbf{k}} + v_{\mathbf{k}}^\beta g_{\mathbf{k}}^\dagger \qquad (2.24)$$

$$\beta_{-\mathbf{k},\downarrow} = v_{\mathbf{k}}^\beta d_{\mathbf{k}}^\dagger - u_{\mathbf{k}}^\beta g_{\mathbf{k}} \qquad (2.25)$$

The diagonalized mean-field Hamiltonian in the superconducting state is

$$H = \sum_{\mathbf{p}}{}' \left[\Omega_{\mathbf{k}}^\alpha (a_{\mathbf{k}}^\dagger a_{\mathbf{k}} + b_{\mathbf{k}}^\dagger b_{\mathbf{k}}) + \Omega_{\mathbf{k}}^\beta (d_{\mathbf{k}}^\dagger d_{\mathbf{k}} + g_{\mathbf{k}}^\dagger g_{\mathbf{k}}) \right] \qquad (2.26)$$

where the energies of the Bogoliubov quasiparticle excitations are given by

$$\Omega_{\mathbf{k}}^{\alpha,\beta} = \sqrt{\left(E_{\mathbf{k}}^{\alpha,\beta}\right)^2 + \left(\Delta_{\mathbf{k}}^{\alpha,\beta}\right)^2} \qquad (2.27)$$

2.5 Theory of Heavy Fermion QPI

Now that the theoretical form of the electronic structure for the superconducting state of a heavy fermion material has been determined, it remains to connect this to the experimentally observed QPI, which describes the scattering between states rather than the states themselves. This theory was developed for URu_2Si_2 in Ref. [27], along the following lines. The QPI spectrum is the power spectrum determined from the Fourier transform of the real space differential conductance, dI/dV. In a heavy fermion material, the presence of the STM tip introduces a tunneling Hamiltonian

$$H_T = -\sum_{\mathbf{r},\sigma} \left(t_c c_{\mathbf{r},\sigma}^\dagger d_\sigma + t_f f_{\mathbf{r},\sigma}^\dagger d_\sigma + H.c. \right) \qquad (2.28)$$

2.5 Theory of Heavy Fermion QPI

with the operator d_σ annihilating an electron in the lead with spin σ, t_c is the hopping between the tip and the conduction band, and $t_f = t_{f,0} r_0$ is the hopping between the tip and f-electrons, renormalized by the expectation value of the slave boson. To account for the complexity of various tunneling processes between the tip and the system it is helpful to introduce a matrix notation. The Green's functions of Eqs. (2.8)–(2.10) (now in real space) can be combined into compact matrix form:

$$\hat{G}_\sigma(\mathbf{r},\mathbf{r},E) = \begin{pmatrix} G_{cc}(\mathbf{r},\mathbf{r},\sigma,E) & G_{cf}(\mathbf{r},\mathbf{r},\sigma,E) \\ G_{fc}(\mathbf{r},\mathbf{r},\sigma,E) & G_{ff}(\mathbf{r},\mathbf{r},\sigma,E) \end{pmatrix} \quad (2.29)$$

with $G_{fc}(\mathbf{r},\mathbf{r},\sigma,E) = G_{cf}(\mathbf{r},\mathbf{r},\sigma,E)$, as can be shown diagrammatically. The most general expression for the tunneling current utilizes the Keldysh formalism (employed in Chaps. 5 and 6). However, in the limit of weak tip-system coupling $(t_c, t_f)\, dI/dV$ one may derive an expression that contains terms proportional to the densities of states of the c and f-electrons, but also a quantum interference term:

$$\frac{dI(\mathbf{r},E)}{dV} = \frac{\pi e}{\hbar} N_t \sum_{i,j=1}^{2} [\hat{t}\,\mathrm{Im}\hat{G}(\mathbf{r},\mathbf{r},E)\hat{t}]_{ij}$$

$$= \frac{2\pi e}{\hbar} N_t \left[t_c^2 N_c(\mathbf{r},E) + t_f^2 N_f(\mathbf{r},E) + t_c t_f N_{cf}(\mathbf{r},E) + t_f t_c N_{fc}(\mathbf{r},E) \right] \quad (2.30)$$

in this expression the tip-system hopping matrix is defined via

$$\hat{t} = \begin{pmatrix} -t_c & 0 \\ 0 & -t_f \end{pmatrix} \quad (2.31)$$

and N_t is the tip density of states. Equation (2.30) can be understood pictorially in terms of the multiple tunneling paths between the STM tip and the sample. For instance, the term with $i = j = 1$ represents an electron hopping between the tip and the conduction band, whereas $i = j = 2$ is the same for the f-band. The off-diagonal terms, on the other hand, represent hopping from the tip to either the c- or f-band and returning from the other one. As discussed above in Sect. 2.1, QPI measures the oscillations in the density of states due to the scattering of electrons off defects. To incorporate defect scattering we may introduce the Born (first-order) scattering approximation, valid in the dilute limit of defect concentration. Experimentally this will be accessible if the STM is positioned over a relatively clean portion of the sample. The expression for the Fourier-transformed QPI signal is given by

$$\bar{g}(\mathbf{q},e) \equiv \frac{dI(\mathbf{q},e)}{dV} = \frac{\pi e}{\hbar} N_t \sum_{i,j=1}^{2} [\hat{t}\hat{N}(\mathbf{q},E)\hat{t}]_{ij} \quad (2.32)$$

where the factor

$$\hat{N}(\mathbf{q},E) = -\frac{1}{\pi}\mathrm{Im}\int \frac{d^2k}{(2\pi)^2} \hat{G}(\mathbf{k},E)\hat{U}\hat{G}(\mathbf{k}+\mathbf{q},E) \quad (2.33)$$

now contains the effect of impurity scattering as encoded in

$$\hat{U} = \begin{pmatrix} U_{cc} & U_{cf} \\ U_{fc} & U_{ff} \end{pmatrix} \quad (2.34)$$

Here U_{cc} and U_{ff} are the potentials for scattering in the c- and f-bands, respectively, whereas U_{fc} and U_{cf} scatter electrons between the bands. Equations (2.32)–(2.34) are the expressions ultimately needed to model the experimentally determined scattering band structure of Fig. 2.3. The practical task then is to identify functional forms of $\varepsilon_{\mathbf{k}}^c$, $\varepsilon_{\mathbf{k}}^f$, $s_{\mathbf{k}}^f$, and the values of U_{cc}, U_{cf}, and U_{ff} that can be used to reproduce the experimental QPI results. We note that the previous work on QPI in CeCoIn5 [28] considered only a single band, and thereby neglected the possibility of interference between different tunneling paths.

Notice that the equations used here to model the overall features of the QPI do not depend on the superconducting properties. It is to be expected that, given the low critical temperature of $T_c = 2.3$ K, the magnitudes of the superconducting gaps $\Delta_{\mathbf{k}}^{\alpha,\beta}$ will also be very small (as follows from the proportionality of Δ and T_c in BCS theory). The superconductivity, therefore, will only modify the differential conductance, and hence the QPI, very close to the Fermi energy. For data on the scale of Fig. 2.3, the normal state properties (of the low temperature heavy Fermi liquid phase) will suffice to explain the QPI. Later we will consider the QPI very close to the Fermi level, and the additional effects of the superconducting gap will need to be included.

2.6 CeCoIn5 QPI at Large Energies

To better understand the complex evolution of the heavy fermion electronic structure, as revealed in the QPI experiments, we begin with a schematic picture of the low-energy band dispersions. Figure 2.4a shows the situation at temperatures above the crossover to the hybridized Kondo-screened state, T_K. The conduction electron band (black line in Fig. 2.4a) is highly dispersive, in contrast to the narrow f-electron band (red line), which has a bandwidth less than 20 meV. Below T_K, the two bands hybridize due to the Kondo screening of the f-electron moments by the conduction electrons, forming heavy quasiparticle bands α and β with an avoided crossing (blue lines in Fig. 2.4b). At still lower temperatures, $T < T_c$, the system becomes superconducting, with a gap opening at the Fermi surfaces of the α- and β-bands. This phenomenon is indicated by the orange lines in Fig. 2.4b.

We now seek explicit functional forms for the conduction and f-electron dispersions, $\varepsilon_{\mathbf{k}}^c$ and $\varepsilon_{\mathbf{k}}^f$, respectively. To do this, we employ standard tight-binding expressions for nearest-neighbor, next-nearest, etc. hopping of electrons between sites. One determines the band structure by requiring that the theoretically calculated QPI (using a given set of parameters) reproduce the experimentally observed spectrum. The QPI data far away from the crossing point of the unhybridized bands

2.6 CeCoIn5 QPI at Large Energies

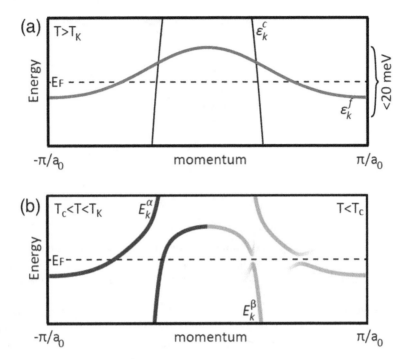

Fig. 2.4 Schematic band structure for a heavy fermion system in the temperature regimes (**a**) $T_c < T_K < T$ where the conduction and f-electrons are unhybridized and (**b**) $T_c < T < T_K$, the hybridized Kondo-screened state (blue line) and $T < T_c < T_K$, the superconducting state (orange line)

reflect the electronic structure at temperatures above T_K. This allows one to fit the individual dispersions $\varepsilon_\mathbf{k}^c$ and $\varepsilon_\mathbf{k}^f$ for the conduction and f-electrons, respectively. It is found that nearest-neighbor hoppings alone do not suffice for obtaining good agreement. Additional further-than-nearest hoppings were included to improve the agreement, leading to the following dispersions [5, 29]:

$$\varepsilon_\mathbf{k}^c = -2t_{c1}\left[\cos(k_x) + \cos(k_y)\right] - 4t_{c2}\cos(k_x)\cos(k_y) \\ - 2t_{c3}\left[\cos(2k_x) + \cos(2k_y)\right] - \mu \quad (2.35)$$

$$\varepsilon_\mathbf{k}^f = -2t_{f1}\left[\cos(k_x) + \cos(k_y)\right] - 4t_{f2}\cos(k_x)\cos(k_y) \\ - 2t_{f3}\left[\cos(2k_x) + \cos(2k_y)\right] \\ - 4t_{f5}\cos(2k_x)\cos(2k_y) - 2t_{f7}\left[\cos(3k_x) + \cos(3k_y)\right] + \varepsilon_f \quad (2.36)$$

with the parameters given in Table 2.1, where the spatial forms of the hoppings $t_{\mathbf{r}-\mathbf{r}'}$ are identified in the second column by $\mathbf{r} - \mathbf{r}' = (r_x - r'_x, r_y - r'_y)$. Note that two slightly different sets of parameters were used in Refs. [5] and [29]. In the former

Table 2.1 Tight-binding parameters for c and f electron dispersions of CeCoIn$_5$

Variable	Distance	Ref. [5] (meV)	Ref. [29] (meV)
t_{c1}	$(\pm 1, 0)$ or $(0, \pm 1)$	-50.0	-50.0
t_{c2}	$(\pm 1, \pm 1)$	-13.34	-13.36
t_{c3}	$(\pm 2, 0)$ or $(0, \pm 2)$	16.7	16.73
μ_c	–	151.51	151.51
t_{f1}	$(\pm 1, 0)$ or $(0, \pm 1)$	-0.85	-0.85
t_{f2}	$(\pm 1, \pm 1)$	-0.45	-0.35
t_{f3}	$(\pm 2, 0)$ or $(0, \pm 2)$	-0.7	-0.8
t_{f5}	$(\pm 2, \pm 2)$	0.125	0.1
t_{f7}	$(\pm 3, 0)$ or $(0, \pm 3)$	0.15	0.09
ε_f	–	0.5	0.5

case, the band structure was determined entirely on the basis of comparison with the experimental QPI results. In the latter case, the set of parameters was slightly adjusted in order to obtain good agreement with several other experiments as well (Sect. 3.4.1).

The QPI data near the avoided crossing of the heavy quasiparticle bands is influenced by the hybridization, $s_\mathbf{k}$, between the conduction and f-electrons. Again requiring that the theoretically calculated QPI reproduce the experimental results leads to the good fit

$$s_\mathbf{k} = s_0 + s_1 [\sin(k_x) \sin(k_y)]^2 \qquad (2.37)$$

where $s_0 = 3.0$ meV and $s_1 = 7.0$ meV. The resulting dispersions and Fermi surfaces (using the parameters of Ref. [5]) are shown in Fig. 2.5. When the QPI is calculated theoretically using Eq. (2.32) and the intensity maxima are extracted and compared with those determined experimentally, the results of Fig. 2.6 are obtained. One notices that the theoretical model reproduces the major branches of the experimental QPI maxima, largely within the experimental uncertainties. Thus, our model allowed us to extract the complex electronic band structure of CeCoIn$_5$ from the experimental QPI results.

2.7 CeCoIn$_5$ QPI at Small Energies

We now turn to the energy range in the immediate vicinity of the Fermi surface, in which the effects of superconductivity become important. An example of the measured superconducting gap in the dI/dV is given in Fig. 2.7. By measuring the magnitude of the gap (2Δ is the distance between the peaks) at each site in a 2D field of view, the Davis group at Cornell produced the gap map shown in Fig. 2.8. Notice the high degree of uniformity in the magnitude of the gap at sites

2.7 CeCoIn$_5$ QPI at Small Energies

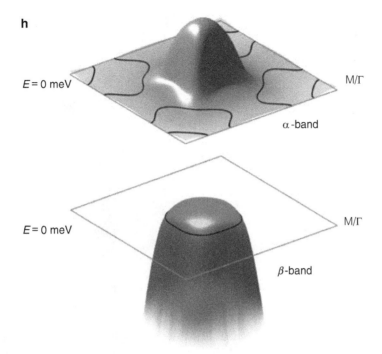

Fig. 2.5 Dispersions of the heavy quasiparticle bands in the theoretical model of CeCoIn$_5$ [5]

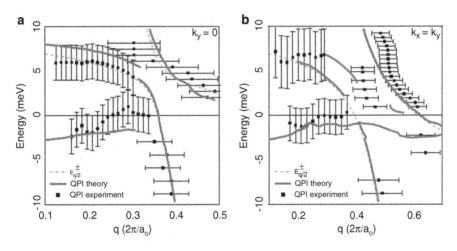

Fig. 2.6 Comparison of theoretical and experimental maxima in the QPI spectrum for one-dimensional cuts along the directions (**a**) $(0, 0) \rightarrow (0, 2\pi/a_0)$ and (**b**) $(0, 0) \rightarrow (2\pi/a_0, 2\pi/a_0)$ [5]

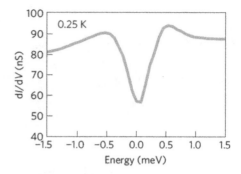

Fig. 2.7 Example of the superconducting gap measured in the differential conductance of CeCoIn$_5$ at T = 0.25 K [5]

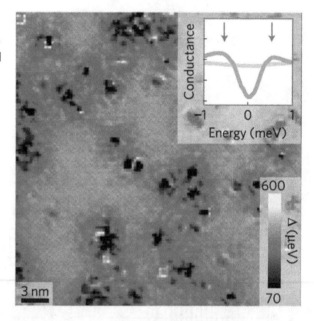

Fig. 2.8 Superconducting gap as a function of spatial position in a 2D field of view on the surface of CeCoIn$_5$ [5]

away from the defects in the lattice, which is a consequence of the high purity attainable in the growth of CeCoIn$_5$. (This property has made CeCoIn$_5$ intriguing as a possible system to realize the long-sought Fulde-Ferrell-Larkin-Ovchinnikov (FFLO) superconducting state at high magnetic fields [30–33]).

A more detailed analysis of the superconducting state can be performed using QPI spectroscopy, which also allows for direct comparison with the theoretical framework of BCS. To this end, we must first extend the theoretical expressions for the QPI signal to include superconductivity. The general structure, Eqs. (2.32) and (2.33), is the same as before, but with the replacements

2.7 CeCoIn₅ QPI at Small Energies

$$\hat{U} = \begin{pmatrix} U_{cc} & U_{cf} & 0 & 0 \\ U_{fc} & U_{ff} & 0 & 0 \\ 0 & 0 & -U_{cc} & -U_{cf} \\ 0 & 0 & -U_{fc} & -U_{ff} \end{pmatrix} \quad (2.38)$$

$$\hat{t} = \begin{pmatrix} -t_c & 0 & 0 & 0 \\ 0 & -t_f & 0 & 0 \\ 0 & 0 & t_c & 0 \\ 0 & 0 & 0 & t_f \end{pmatrix} \quad (2.39)$$

and

$$\hat{G}(\mathbf{k}, E) = \begin{pmatrix} G_{cc}(\mathbf{k}, \sigma, E) & G_{cf}(\mathbf{k}, \sigma, E) & F_{cc}(\mathbf{k}, E) & F_{cf}(\mathbf{k}, E) \\ G_{fc}(\mathbf{k}, \sigma, E) & G_{ff}(\mathbf{k}, \sigma, E) & F_{fc}(\mathbf{k}, E) & F_{ff}(\mathbf{k}, E) \\ F_{cc}(\mathbf{k}, E) & F_{cf}(\mathbf{k}, E) & -G_{cc}(\mathbf{k}, \sigma, -E) & -G_{cf}(\mathbf{k}, \sigma, -E) \\ F_{fc}(\mathbf{k}, E) & F_{ff}(\mathbf{k}, E) & -G_{fc}(\mathbf{k}, \sigma, -E) & -G_{ff}(\mathbf{k}, \sigma, -E) \end{pmatrix} \quad (2.40)$$

Furthermore, the forms of the normal Green's functions ($\gamma, \zeta = c, f$)

$$G_{\gamma,\zeta}(\mathbf{r}, \mathbf{r}, \sigma, \tau) = -\langle T_\tau \gamma_{\mathbf{r}',\sigma}(\tau) \zeta^\dagger_{\mathbf{r},\sigma}(0) \rangle \quad (2.41)$$

are modified in the superconducting state, and new anomalous (or Gor'kov) Green's functions are also introduced according to

$$F_{\gamma,\zeta}(\mathbf{r}, \mathbf{r}, \sigma, \tau) = -\langle T_\tau \gamma^\dagger_{\mathbf{r}',\uparrow}(\tau) \zeta^\dagger_{\mathbf{r},\downarrow}(0) \rangle \quad (2.42)$$

The explicit forms of these Green's functions in the superconducting state of the heavy Fermi liquid are

$$G_{cc}(\mathbf{k}, \sigma, \omega) = w_\mathbf{k}^2 \frac{\omega + i\Gamma + E_\mathbf{k}^\alpha}{(\omega + i\Gamma)^2 - (\Omega_\mathbf{k}^\alpha)^2} + x_\mathbf{k}^2 \frac{\omega + i\Gamma + E_\mathbf{k}^\beta}{(\omega + i\Gamma)^2 - (\Omega_\mathbf{k}^\beta)^2} \quad (2.43)$$

$$G_{cf}(\mathbf{k}, \sigma, \omega) = w_\mathbf{k} x_\mathbf{k} \left[\frac{\omega + i\Gamma + E_\mathbf{k}^\alpha}{(\omega + i\Gamma)^2 - (\Omega_\mathbf{k}^\alpha)^2} - \frac{\omega + i\Gamma + E_\mathbf{k}^\beta}{(\omega + i\Gamma)^2 - (\Omega_\mathbf{k}^\beta)^2} \right] \quad (2.44)$$

$$G_{ff}(\mathbf{k}, \sigma, \omega) = x_\mathbf{k}^2 \frac{\omega + i\Gamma + E_\mathbf{k}^\alpha}{(\omega + i\Gamma)^2 - (\Omega_\mathbf{k}^\alpha)^2} + w_\mathbf{k}^2 \frac{\omega + i\Gamma + E_\mathbf{k}^\beta}{(\omega + i\Gamma)^2 - (\Omega_\mathbf{k}^\beta)^2} \quad (2.45)$$

Fig. 2.9 Superconducting gap along the Fermi surface as determined by fitting the QPI data. The magnitude is indicated by the height of the red line above the xy-plane. The sign of the gap is given by the background color of the Brillouin zone, which for the α-band is yellow for positive and blue for negative gap values (the convention is reversed for the β-band) [5]

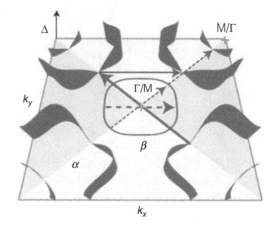

$$F_{cc}(\mathbf{k}, \omega) = w_\mathbf{k}^2 \frac{\Delta_\mathbf{k}^\alpha}{(\omega + i\Gamma)^2 - (\Omega_\mathbf{k}^\alpha)^2} + x_\mathbf{k}^2 \frac{\Delta_\mathbf{k}^\beta}{(\omega + i\Gamma)^2 - (\Omega_\mathbf{k}^\beta)^2} \quad (2.46)$$

$$F_{cf}(\mathbf{k}, \omega) = w_\mathbf{k} x_\mathbf{k} \left[\frac{\Delta_\mathbf{k}^\alpha}{(\omega + i\Gamma)^2 - (\Omega_\mathbf{k}^\alpha)^2} - \frac{\Delta_\mathbf{k}^\beta}{(\omega + i\Gamma)^2 - (\Omega_\mathbf{k}^\beta)^2} \right] \quad (2.47)$$

$$F_{ff}(\mathbf{k}, \omega) = x_\mathbf{k}^2 \frac{\Delta_\mathbf{k}^\alpha}{(\omega + i\Gamma)^2 - (\Omega_\mathbf{k}^\alpha)^2} + w_\mathbf{k}^2 \frac{\Delta_\mathbf{k}^\beta}{(\omega + i\Gamma)^2 - (\Omega_\mathbf{k}^\beta)^2} \quad (2.48)$$

with the heavy fermion coherence factors defined in Eqs. (2.11) and (2.12). With these equations, we may substitute proposed forms of the superconducting gap functions $\Delta_\mathbf{k}^{\alpha,\beta}$ to calculate the QPI theoretically and compare with experiment. First we note that the overall magnitude of the gap is constrained by the tunneling data. We may consider three different scenarios to try to reproduce the experimental results in a simple way. First, we can try $d_{x^2-y^2}$-symmetry gaps on both the α and β bands, but with unequal magnitudes, that is

$$\Delta_\mathbf{k}^{\alpha,\beta} = \frac{\Delta_0^{\alpha,\beta}}{2}(\cos(k_x) - \cos(k_y)) \quad (2.49)$$

with $\Delta_0^\alpha = 1.0$ meV and $\Delta_0^\beta = -0.2$ meV as determined by comparison of the theoretical and experimental QPI results. The gap is shown along the Fermi surface in Fig. 2.9. One should note that with these parameters, the maximum gap on the β-band is $\approx 50\,\mu$eV, which is smaller than the experimental resolution of 75 μeV. This would then explain the lack of an easily visible feature in the tunneling spectra associated with the β-band gap. However, the inclusion of this gap can have consequences for the QPI spectra. Figure 2.10 presents the QPI for the experimental (a–c) and theoretical (d–f) cases (the latter using unequal gaps with $d_{x^2-y^2}$-symmetry). The colored circles indicate key scattering wavelengths in

2.7 CeCoIn$_5$ QPI at Small Energies

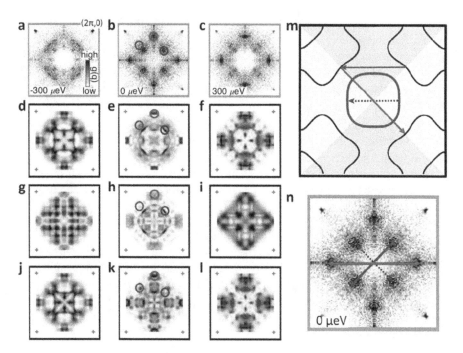

Fig. 2.10 Experimental and theoretical QPI for several possible superconducting gap symmetries. (**a**)–(**c**) Experimental QPI data for the energies −300, 0, and 300 μeV, respectively. (**d**)–(**l**) Theoretical simulations of QPI for a $d_{x^2-y^2}$-symmetry gap [(**d**)–(**f**)], d_{xy}-symmetry gap [(**g**)–(**i**)], and a gap with $d_{x^2-y^2}$-symmetry of equal magnitude on the α- and β-bands [(**j**)–(**l**)]. The red and brown circles indicate the strongest internodal scattering vectors and the blue circles show the strongest β-band scattering vector. These vectors are shown on the Fermi surface in (**m**) and overlaid on the E = 0 meV experimental data in (**n**). Note that only the case (**d**)–(**f**) with $d_{x^2-y^2}$ and unequal gaps reproduces all the important scattering vectors [5]

the spectra, which are well-reproduced by the theoretical model. The corresponding scattering processes across the Fermi surface are shown in panel m using matching colored arrows, and are illustrated on top of the experimental results in panel n (note that the axes of the experimental q-space are rotated by 45° compared to the theoretical Brillouin zone). While theoretical models of QPI spectra are known to successfully reproduce the geometric information contained in experiment, matching the observed intensities is more difficult (although some work has been done in this direction [34]).

We can compare this result with the one obtained under the assumption of a different d-wave symmetry, namely, d_{xy}, for which a basis function is

$$\Delta_{\mathbf{k}}^{\alpha,\beta} = \Delta_0^{\alpha,\beta}(\sin(k_x)\sin(k_y)) \qquad (2.50)$$

Using the values $\Delta_0^\alpha = 1.0$ meV and $\Delta_0^\beta = -0.2$ meV, the resulting QPI patterns are shown in panels (**g**–**i**) of Fig. 2.10. As is readily apparent, this choice of symmetry fails to reproduce the correct QPI at zero energy, notably the features indicated by the red and brown circles in the experimental data are absent in the calculations. Under the assumption of $d_{x^2-y^2}$-symmetry, one might attempt to alter the relative magnitudes of the gaps on the α and β-bands. In doing this, we found that it is necessary that the magnitude of the β-band gap be considerably smaller than that of the α-band. For example, if Δ_0^β is adjusted such that the maximum gaps on the two bands are equal ($\Delta_0^\beta = -2.6\,\text{meV}$), the results in Fig. 2.10j–l are obtained. Unlike the d_{xy}-symmetry, this reproduces the features of the red and brown circles, but it fails to do so for the blue circle. Thus we conclude that unequal-magnitude $d_{x^2-y^2}$-symmetry gaps are the best for reproducing the experimental results. This of course does not preclude other more complicated symmetries but it does limit the space of possibilities considerably.

To conclude, we have shown that it is possible to successfully and quantitatively model the heavy-fermion band structure of CeCoIn$_5$ in the superconducting state. The relevant empirical input was extracted using the QPI measurement technique of STM-STS experiments. The theory provided a concrete theoretical picture and a rationalization of both the high-energy spectra (dominated by the normal state heavy Fermi liquid) and at very low energies, where superconductivity is significant. This paves the way for further joint experimental-theoretical studies on CeCoIn$_5$ and other heavy fermion materials.

References

1. J.R. Schrieffer, *Theory of Superconductivity*, revised edn. (Perseus Books, Reading, 1999)
2. D.J. Scalapino, A common thread: the pairing interaction for unconventional superconductors. Rev. Mod. Phys. **84**(4), 1383–1417 (2012)
3. K. Izawa, H. Yamaguchi, Y. Matsuda, H. Shishido, R. Settai, Y. Onuki, Angular position of nodes in the superconducting gap of quasi-2D heavy-fermion superconductor CeCoIn$_5$. Phys. Rev. Lett. **87**(5), 057002 (2001)
4. H. Aoki, T. Sakakibara, H. Shishido, R. Settai, Y. Ōnuki, P. Miranović, K. Machida, Field-angle dependence of the zero-energy density of states in the unconventional heavy-fermion superconductor CeCoIn$_5$. J. Phys. Condens. Matter **16**(3), L13 (2004)
5. M.P. Allan, F. Massee, D.K. Morr, J. Van Dyke, A.W. Rost, A.P. Mackenzie, C. Petrovic, J.C. Davis, Imaging Cooper pairing of heavy fermions in CeCoIn$_5$. Nat. Phys. **9**(8), 468–473 (2013)
6. B.B. Zhou, S. Misra, E.H. da Silva Neto, P. Aynajian, R.E. Baumbach, J.D. Thompson, E.D. Bauer, A. Yazdani, Visualizing nodal heavy fermion superconductivity in CeCoIn$_5$. Nat. Phys. **9**(8), 474–479 (2013)
7. J.A. Stroscio, R.M. Feenstra, A.P. Fein, Electronic Structure of the Si(111)2×1 Surface by Scanning-Tunneling Microscopy. Phys. Rev. Lett. **57**(20), 2579–2582 (1986)
8. M.F. Crommie, C.P. Lutz, D.M. Eigler, Imaging standing waves in a two-dimensional electron gas. Nature **363**(6429), 524–527 (1993)
9. S.H. Pan, J.P. O'Neal, R.L. Badzey, C. Chamon, H. Ding, J.R. Engelbrecht, Z. Wang, H. Eisaki, S. Uchida, A.K. Gupta, K.-W. Ng, E.W. Hudson, K.M. Lang, J.C. Davis, Microscopic electronic inhomogeneity in the high-Tc superconductor Bi$_2$Sr$_2$CaCu$_2$O$_{8+x}$. Nature **413**(6853), 282–285 (2001)

10. J. Tersoff, D.R. Hamann, Theory and application for the scanning tunneling microscope. Phys. Rev. Lett. **50**(25), 1998–2001 (1983)
11. A.L. de Lozanne, S.A. Elrod, C.F. Quate, Spatial variations in the superconductivity of Nb_3Sn measured by low-temperature tunneling microscopy. Phys. Rev. Lett. **54**(22), 2433–2436 (1985)
12. M.H. Hamidian, A.R. Schmidt, I.A. Firmo, M.P. Allan, P. Bradley, J.D. Garrett, T.J. Williams, G.M. Luke, Y. Dubi, A.V. Balatsky, J.C. Davis, How Kondo-holes create intense nanoscale heavy-fermion hybridization disorder. Proc. Natl. Acad. Sci. **108**(45), 18233–18237 (2011)
13. Y. Hasegawa, P. Avouris, Direct observation of standing wave formation at surface steps using scanning tunneling spectroscopy. Phys. Rev. Lett. **71**(7), 1071–1074 (1993)
14. L. Petersen, P.T. Sprunger, P. Hofmann, E. Lægsgaard, B.G. Briner, M. Doering, H.-P. Rust, A.M. Bradshaw, F. Besenbacher, E.W. Plummer, Direct imaging of the two-dimensional Fermi contour: Fourier-transform STM. Phys. Rev. B **57**(12), R6858–R6861 (1998)
15. K. Fujita, M. Hamidian, I. Firmo, S. Mukhopadhyay, C.K. Kim, H. Eisaki, S.-i. Uchida, J.C. Davis, Spectroscopic imaging STM: atomic-scale visualization of electronic structure and symmetry in underdoped cuprates, in *Strongly Correlated Systems*, ed. by A. Avella, F. Mancini. Springer Series in Solid-State Sciences, vol. 180 (Springer, Berlin, 2015), pp. 73–109. https://doi.org/10.1007/978-3-662-44133-6_3
16. G.D. Mahan, *Many-Particle Physics*, 3rd edn., 2000 edn. (Springer, New York, 2000)
17. J. Friedel, Metallic alloys. Il Nuovo Cimento **7**(2), 287–311 (1958)
18. K. Andres, J.E. Graebner, H.R. Ott, $4f$-virtual-bound-state formation in $CeAl_3$ at low temperatures. Phys. Rev. Lett. **35**(26), 1779–1782 (1975)
19. N. Read, D.M. Newns, On the solution of the Coqblin-Schreiffer Hamiltonian by the large-N expansion technique. J. Phys. C Solid State Phys. **16**(17), 3273 (1983)
20. D.M. Newns, N. Read, Mean-field theory of intermediate valence/heavy fermion systems. Adv. Phys. **36**(6), 799–849 (1987)
21. P. Coleman, New approach to the mixed-valence problem. Phys. Rev. B **29**(6), 3035–3044 (1984)
22. A.J. Millis, P.A. Lee, Large-orbital-degeneracy expansion for the lattice Anderson model. Phys. Rev. B **35**(7), 3394–3414 (1987)
23. M.T. Béal-Monod, C. Bourbonnais, V.J. Emery, Possible superconductivity in nearly antiferromagnetic itinerant fermion systems. Phys. Rev. B **34**(11), 7716–7720 (1986)
24. K. Miyake, S. Schmitt-Rink, C.M. Varma, Spin-fluctuation-mediated even-parity pairing in heavy-fermion superconductors. Phys. Rev. B **34**(9), 6554–6556 (1986)
25. M. Lavagna, A.J. Millis, P.A. Lee, d-wave superconductivity in the large-degeneracy limit of the Anderson lattice. Phys. Rev. Lett. **58**(3), 266–269 (1987)
26. R. Flint, P. Coleman, Tandem pairing in heavy-fermion superconductors. Phys. Rev. Lett. **105**(24), 246404 (2010)
27. T. Yuan, J. Figgins, D.K. Morr, Hidden order transition in URu_2Si_2: evidence for the emergence of a coherent Anderson lattice from scanning tunneling spectroscopy. Phys. Rev. B **86**(3), 035129 (2012)
28. A. Akbari, P. Thalmeier, I. Eremin, Quasiparticle interference in the heavy-fermion superconductor $CeCoIn_5$. Phys. Rev. B **84**(13), 134505 (2011)
29. J.S. Van Dyke, F. Massee, M.P. Allan, J.C.S. Davis, C. Petrovic, D.K. Morr, Direct evidence for a magnetic f-electron–mediated pairing mechanism of heavy-fermion superconductivity in $CeCoIn_5$. Proc. Natl. Acad. Sci. **111**(32), 11663–11667 (2014)
30. A. Bianchi, R. Movshovich, C. Capan, P.G. Pagliuso, J.L. Sarrao, Possible Fulde-Ferrell-Larkin-Ovchinnikov superconducting state in $CeCoIn_5$. Phys. Rev. Lett. **91**(18), 187004 (2003)
31. H. Won, K. Maki, S. Haas, N. Oeschler, F. Weickert, P. Gegenwart, Upper critical field and Fulde-Ferrell-Larkin-Ovchinnikov state in $CeCoIn_5$. Phys. Rev. B **69**(18), 180504 (2004)
32. C.F. Miclea, M. Nicklas, D. Parker, K. Maki, J.L. Sarrao, J.D. Thompson, G. Sparn, F. Steglich, Pressure dependence of the Fulde-Ferrell-Larkin-Ovchinnikov state in $CeCoIn_5$. Phys. Rev. Lett. **96**(11), 1–4 (2006)

33. K. Kumagai, M. Saitoh, T. Oyaizu, Y. Furukawa, S. Takashima, M. Nohara, H. Takagi, Y. Matsuda, Fulde-Ferrell-Larkin-Ovchinnikov state in a perpendicular field of quasi-two-dimensional $CeCoIn_5$. Phys. Rev. Lett. **97**(22), 227002 (2006)
34. F.P. Toldin, J. Figgins, S. Kirchner, D.K. Morr, Disorder and quasiparticle interference in heavy-fermion materials. Phys. Rev. B **88**(8), 081101 (2013)

Chapter 3
Pairing Mechanism in CeCoIn$_5$

3.1 Heavy Fermion Superconductivity

Although superconductivity was discovered in CeCu$_2$Si$_2$ as early as 1979 [1], it took some time to overcome the conventional wisdom that magnetism was necessarily detrimental to Cooper pairing. At present there are several dozen known heavy fermion superconductors, which display a wide range of behaviors. UPt$_3$ possesses multiple superconducting phases with different symmetries, whereas CeMIn$_5$ (M=Co,Ir,Rh), the so-called "115 materials," have complex phase diagrams that include superconductivity, antiferromagnetism, and non-Fermi liquid behavior [2, 3].

A number of superconducting pairing mechanisms have been proposed to provide the attractive force binding the Cooper pairs. The most widely accepted mechanism is spin fluctuations between the heavy quasiparticles in the low temperature Fermi liquid state. A notable feature of this mechanism is its implications for the superconducting gap symmetry. Assuming total rotational invariance of the system (neglecting the underlying crystal structure of the lattice), spin fluctuations were found to suppress both singlet and triplet pairing [4]. However, including the crystal structure in the calculation reveals that the system can partially avoid the electron repulsion by establishing an anisotropic superconducting gap (generally of d-wave symmetry) [5–8]. Other more complex mechanisms have also been proposed [9, 10]. In particular, recent work suggests the possibility of composite pairing between conduction electrons and local moments [11–14] as an explanation of fully gapped superconductivity in Yb-doped CeCoIn$_5$ [15].

The quantitative description of superconductivity in any system requires two key pieces of information. First, one must obtain the detailed low energy band structure in the normal state, out of which superconductivity emerges. Second, one needs the microscopic form of the pairing interaction responsible for the formation of Cooper pairs and the resulting well-known phenomena observed in the superconducting state. The QPI experiments analyzed in the previous chapter provide us with precisely the information needed for a quantitative study of superconductivity in CeCoIn$_5$, which has never before been achieved. In the following we discuss how the magnetic interaction f-electrons, which gives rise to the curvature of the f-bands, may be determined from the QPI data relevant for the normal state. Under the assumption that this same f-electron interaction also provides for the Cooper pairing, we derive a series of predictions about the superconducting state of CeCoIn$_5$, in good agreement with experiment.

3.2 Extraction of the Magnetic Interaction

In the model proposed in Eq. (2.3) the interaction between the localized magnetic f-electrons is captured in the Heisenberg term, Eq. (3.1).

$$H_H = \sum_{\mathbf{r},\mathbf{r}'} I_{\mathbf{r},\mathbf{r}'} \mathbf{S}_\mathbf{r} \cdot \mathbf{S}_{\mathbf{r}'} \tag{3.1}$$

This term was decoupled in Chap. 2 to give a dispersion to the f-electrons, via the self-consistency equation (2.6). In Chap. 2, this dispersion was obtained by comparison with the experimental QPI and encoded in the hopping parameters t_{f1}–t_{f7} and ε_f of Table 2.1. The curvature of the f-band directly arises from magnetic interaction $I(\mathbf{r}, \mathbf{r}')$ treated within the mean-field approximation. Inverting the self-consistency equation to solve for $I(\mathbf{r}, \mathbf{r}')$ in terms of $t_f(\mathbf{r}, \mathbf{r}')$ allows one to quantitatively determine the form of the magnetic interaction:

$$I_{\mathbf{r},\mathbf{r}'} = -\frac{\pi t_f(\mathbf{r}, \mathbf{r}')}{\int_{-\infty}^{\infty} d\omega n_F(\omega) \mathrm{Im} G_{ff}(\mathbf{r}, \mathbf{r}', \omega)} \tag{3.2}$$

$I_{\mathbf{r},\mathbf{r}'}$ will be non-zero only if $t_f(\mathbf{r}, \mathbf{r}')$ is so, which for the band structure extracted from experiment (Table 2.1) is true for t_{f1}–t_{f3}, t_{f5}, and t_{f7}. Using the Green's functions of Eq. (2.9) one may solve Eq. (3.2) for the corresponding values of I, thereby obtaining the form of the underlying magnetic interaction between the f-electrons in Eq. (3.3).

$$I(\mathbf{q}) = 2I_1[\cos(q_x) + \cos(q_y)] + 4I_2[\cos(q_x)\cos(q_y)] + 2I_3[\cos(2q_x) + \cos(2q_y)]$$
$$+ 4I_5[\cos(2q_x)\cos(2q_y)] + 2I_7[\cos(3q_x) + \cos(3q_y)] \tag{3.3}$$

The numerical value of this interaction, employing the experimentally determined parameters of [16], is shown in real space in Fig. 3.1. The numerical values of the interactions (between nearest, next-nearest, etc. neighbors) are given in Table 3.1. Positive values of $I(\mathbf{r} - \mathbf{r}')$ denote antiferromagnetic interactions, whereas negative ones imply ferromagnetic correlations.

3.2 Extraction of the Magnetic Interaction

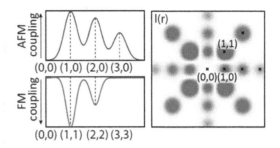

Fig. 3.1 Real space form of the magnetic interaction between f-electrons as extracted from the bandstructure fit to the experiment QPI data [16]

Table 3.1 f-electron magnetic interaction parameters extracted from bandstructure fits to QPI experiments on CeCoIn$_5$

Variable	Distance	Value (meV)
I_1	$(\pm 1, 0)$ or $(0, \pm 1)$	6.44
I_2	$(\pm 1, \pm 1)$	-20.30
I_3	$(\pm 2, 0)$ or $(0, \pm 2)$	6.04
I_5	$(\pm 2, \pm 2)$	-9.65
I_7	$(\pm 3, 0)$ or $(0, \pm 3)$	2.58

We now explore the possibility that the same magnetic interaction that produces the curvature of the f-electron band is also the superconducting pairing interaction. We recall first that the experiments determining the dispersion were in fact done in the superconducting state. By using the self-consistency equation in the normal state, Eq. (2.6) to extract $I(\mathbf{q})$, we in fact neglected the feedback effects of superconductivity on the interaction, an assumption which will be shown reasonable below. The neglect of feedback is important for making the calculations computationally manageable.

We proceed by considering the spin-flip part of the Heisenberg term of model (2.3), written in terms of the Abrikosov pseudofermion representation as

$$H_{sf} = \frac{1}{2N} \sum_{\mathbf{k},\mathbf{p},\mathbf{q}} I(\mathbf{q}) f^{\dagger}_{\mathbf{k}+\mathbf{q},\uparrow} f_{\mathbf{k},\downarrow} f^{\dagger}_{\mathbf{p}-\mathbf{q},\downarrow} f_{\mathbf{p},\uparrow} \qquad (3.4)$$

In the heavy Fermi liquid state, the appropriate degrees of freedom are not c and f-electrons but the heavy α and β-band quasiparticles. Thus, we first transform H_{sf} using the canonical transformations of Eqs. (2.16) and (2.17), followed by a decoupling in the singlet particle–particle channels $\langle \alpha^{\dagger}_{\mathbf{k},\uparrow} \alpha^{\dagger}_{-\mathbf{k},\downarrow} \rangle$ and $\langle \beta^{\dagger}_{\mathbf{k},\uparrow} \beta^{\dagger}_{-\mathbf{k},\downarrow} \rangle$, while neglecting interband pairing. This is justified due to "Fermi surface mismatch": it is not possible to pair electrons between the α- and β-Fermi surfaces in such a way that their total momentum is zero (the lowest energy state in equilibrium).

The superconducting gap functions $\Delta^{\alpha,\beta}_{\mathbf{k}}$, which were determined phenomenologically in Chap. 2, can now be explicitly computed since the microscopic form of the pairing interaction is known. This is the first time such a calculation was achieved for a heavy fermion superconductor. Following the standard BCS mean-field theory, the gaps obey Eqs. (3.5) and (3.6).

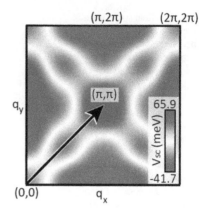

Fig. 3.2 Momentum space form of the magnetic interaction between f-electrons as extracted from the bandstructure fit to the experiment QPI data [16]

$$\Delta_{\mathbf{k}}^{\alpha} = -\frac{x_{\mathbf{k}}^2}{N} \sum_{\mathbf{p}}{}' V_{SC}(\mathbf{p}-\mathbf{k})(x_{\mathbf{p}}^2 \langle \alpha_{\mathbf{p},\uparrow}^{\dagger} \alpha_{-\mathbf{p},\downarrow}^{\dagger} \rangle + w_{\mathbf{p}}^2 \langle \beta_{\mathbf{p},\uparrow}^{\dagger} \beta_{-\mathbf{p},\downarrow}^{\dagger} \rangle) \quad (3.5)$$

$$\Delta_{\mathbf{k}}^{\beta} = -\frac{w_{\mathbf{k}}^2}{N} \sum_{\mathbf{p}}{}' V_{SC}(\mathbf{p}-\mathbf{k})(x_{\mathbf{p}}^2 \langle \alpha_{\mathbf{p},\uparrow}^{\dagger} \alpha_{-\mathbf{p},\downarrow}^{\dagger} \rangle + w_{\mathbf{p}}^2 \langle \beta_{\mathbf{p},\uparrow}^{\dagger} \beta_{-\mathbf{p},\downarrow}^{\dagger} \rangle) \quad (3.6)$$

in which N is the number of sites and the pairing interaction $V_{SC}(\mathbf{q}) = -I(\mathbf{q})/2$ is shown in Fig. 3.2. Note the appearance of the heavy fermion coherence factors $x_{\mathbf{k}}$ and $w_{\mathbf{k}}$ as a result of the canonical transformation from c and f operators to α and β operators. The form of the magnetic interaction $I(\mathbf{q})/2$, peaked as it is at $Q = (\pi, \pi)$, justifies the lack of interband pairing in Eq. (2.19). The mismatch of the α- and β-Fermi surfaces (see Fig. 2.10) implies that scattering between the two occurs only for momenta transfers away from Q where the pairing interaction is weak.

We thus arrive at the same SC mean-field Hamiltonian as was introduced phenomenologically in Eq. (2.19) and diagonalized using the Bogoliubov transformations (2.22)–(2.25) to yield (2.26). Now we apply these transformations to the BCS gap equations as well, finding

$$\Delta_{\mathbf{k}}^{\alpha} = -\frac{x_{\mathbf{k}}^2}{N} \sum_{\mathbf{p}}{}' V_{SC}(\mathbf{p}-\mathbf{k}) \left[x_{\mathbf{p}}^2 \frac{\Delta_{\mathbf{p}}^{\alpha}}{2\Omega_{\mathbf{p}}^{\alpha}} \tanh\left(\frac{\Omega_{\mathbf{p}}^{\alpha}}{2k_B T}\right) + w_{\mathbf{p}}^2 \frac{\Delta_{\mathbf{p}}^{\beta}}{2\Omega_{\mathbf{p}}^{\beta}} \tanh\left(\frac{\Omega_{\mathbf{p}}^{\beta}}{2k_B T}\right) \right] \quad (3.7)$$

$$\Delta_{\mathbf{k}}^{\beta} = -\frac{w_{\mathbf{k}}^2}{N} \sum_{\mathbf{p}}{}' V_{SC}(\mathbf{p}-\mathbf{k}) \left[x_{\mathbf{p}}^2 \frac{\Delta_{\mathbf{p}}^{\alpha}}{2\Omega_{\mathbf{p}}^{\alpha}} \tanh\left(\frac{\Omega_{\mathbf{p}}^{\alpha}}{2k_B T}\right) + w_{\mathbf{p}}^2 \frac{\Delta_{\mathbf{p}}^{\beta}}{2\Omega_{\mathbf{p}}^{\beta}} \tanh\left(\frac{\Omega_{\mathbf{p}}^{\beta}}{2k_B T}\right) \right] \quad (3.8)$$

The heavy fermion coherence factors remain in this form of the gap equation as well, reflecting the fact that the underlying interaction arises from the magnetic f-electrons alone.

3.2 Extraction of the Magnetic Interaction

The determination of the gap symmetry may be performed with the linearized form of the gap equation, which is valid at temperatures near the transition T_c where $\Delta^{\alpha,\beta}$ are small. This results in a simple eigenvalue-eigenvector equation which is computationally more tractable than the full non-linear gap equation. Discretizing the Brillouin zone for **k**-points with $|E_\mathbf{k}^{\alpha,\beta}| \leq \omega_D$ we have

$$\hat{\Delta} = -\hat{V}_{SC}\hat{\Delta} \qquad (3.9)$$

$$(\hat{V}_{SC})_{ij} = \frac{\xi_i^2}{N} V_{SC}(\mathbf{k}_j - \mathbf{k}_i) \left[\frac{\xi_j^2}{2|E_j|} \tanh\left(\frac{|E_j|}{2k_B T}\right)\right] \qquad (3.10)$$

$$\hat{\Delta} = \begin{pmatrix} \Delta_{\mathbf{k}_1}^\alpha \\ \vdots \\ \Delta_{\mathbf{k}_{N_\alpha}}^\alpha \\ \Delta_{\mathbf{k}_1}^\beta \\ \vdots \\ \Delta_{\mathbf{k}_{N_\beta}}^\beta \end{pmatrix} \qquad (3.11)$$

$$\xi_i^2 = \begin{cases} x_{\mathbf{k}_i}^2, & 1 \leq i \leq N_\alpha \\ w_{\mathbf{k}_i}^2, & N_\alpha + 1 \leq i \leq N_\alpha + N_\beta \end{cases} \qquad (3.12)$$

$$E_i = \begin{cases} E_{\mathbf{k}_i}^\alpha, & 1 \leq i \leq N_\alpha \\ E_{\mathbf{k}_i}^\beta, & N_\alpha + 1 \leq i \leq N_\alpha + N_\beta \end{cases} \qquad (3.13)$$

Solving this equation for the eigenvectors $\hat{\Delta}$ allows for the determination of the gap symmetry by direct inspection. Performing the calculation, we find that the SC gaps in both the α- and β-bands possess $d_{x^2-y^2}$-symmetry (shown in Fig. 3.3 for the $T = 0$ gap). Thus, rather than explicitly assuming the gap symmetry (as is often done in theoretical work), we are able to reproduce the correct symmetry of the superconducting gap using calculations based entirely on the normal state band structure and the assumption of a magnetic f-electron pairing interaction. As discussed in Chap. 2, the use of $d_{x^2-y^2}$-symmetry in the theoretical calculation of the QPI uniquely reproduces the important features of the experiments. However, the usual type of QPI experiment (as are the experiments of Chap. 2) is not sensitive to the phase of the superconducting gap. In Sect. 3.3 we discuss a set of phase-sensitive QPI experiments that allows one to distinguish a sign-changing $d_{x^2-y^2}$-symmetry gap and a gap that has nodes but does not change sign along the Fermi surface. Typically, the symmetry of the gap close to T_c does not change as the temperature is lowered towards $T = 0$ (UPt$_3$ is a notable exception). In subsequent calculations of the full momentum dependence of the superconducting gap we assume the correct

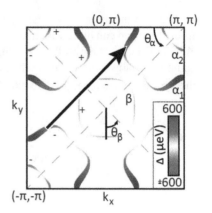

Fig. 3.3 Superconducting gap structure shown along the Fermi surface, as calculated from magnetic f-electron pairing interaction. The black arrow shows points connected by the antiferromagnetic ordering vector $Q = (\pi, \pi)$, between which the gap changes sign [16]

symmetry to reduce the complexity of the problem. That is, the gap equations (3.7) and (3.8) at $T = 0$ are solved for momentum-space points along the Fermi surface in only one eighth of the Brillouin zone, $k_x \geq k_y \geq 0$, since the gap everywhere else is related to this gap by symmetry.

A simple argument for the $d_{x^2-y^2}$-wave pairing symmetry follows from the real space structure of the pairing interaction, Fig. 3.1. Along the bond directions, the antiferromagnetic couplings I_1, I_3, and I_7 result in attractive pairing potentials $V_{SC}(\mathbf{r}) = -I(\mathbf{r})/2 < 0$, between two anti-parallel spins (we assume spin-singlet Cooper pairing). On the other hand, the ferromagnetic I_2 and I_5 along the diagonals are repulsive ($V_{SC} > 0$). Thus, the electrons comprising the pairs can minimize their energy by forming nodes in the Cooper pair wavefunction along the diagonal directions in real space, where the repulsive interaction would otherwise raise the pair's total energy. Alternatively, one sees that in momentum-space the large repulsive interaction near the antiferromagnetic wavevector $Q = (\pi, \pi)$ implies from the BCS gap equation that the gap changes sign between points on the Fermi surface connected by Q, as shown by the black arrow in Fig. 3.3.

We may now solve the full non-linear gap equations (3.7) and (3.8) numerically, so as to obtain the momentum space form of the superconducting gaps on the α and β-bands. The summations are performed for all states with energies satisfying $|E_k^{\alpha,\beta}| \leq \omega_D$, with ω_D a cutoff energy for the spin-fluctuation pairing. The maximum of the gap is proportional to ω_D (as was checked numerically), and so we adjust it to reproduce the experimentally observed maximum of ≈ 0.6 meV. The calculation produces gaps on the α and β-bands with the momentum space structure shown in Fig. 3.4.

We further note that these gap functions can be well fit using the parameterizations of Eqs. (3.14) and (3.15).

$$\Delta_{\mathbf{k}}^\alpha = \frac{\Delta_0^\alpha}{2}\{[\cos(k_x) - \cos(k_y)] + \alpha_1[\cos(2k_x) - \cos(2k_y)] + \alpha_2[\cos(3k_x) - \cos(3k_y)]\} \qquad (3.14)$$

$$\Delta_{\mathbf{k}}^\beta = \frac{\Delta_0^\beta}{2}[\cos(k_x) - \cos(k_y)]^3 \qquad (3.15)$$

3.2 Extraction of the Magnetic Interaction

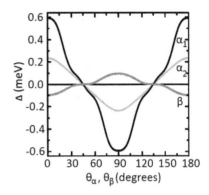

Fig. 3.4 Superconducting gap structure as a function of Fermi surface angle calculated from magnetic f-electron pairing interaction. Fermi surface angle is defined in Fig. 3.3 [16]

with $\Delta_0^\alpha = 0.492$ meV, $\alpha_1 = -0.607$, $\alpha_2 = -0.082$, $\Delta_0^\beta = -1.040$ meV, and $\omega_D = 0.66$ meV was used.

One may also use the linearized gap equation for determining the transition temperature by adjusting the temperature T in Eq. (3.10) such that the linearized equation is satisfied with a maximum eigenvalue $\lambda = 1$. Performing this calculation, we find that $T_c = 2.96$ K, assuming that the quasiparticle lifetimes are infinite ($\Gamma = 0^+$). In real experiments, various sources of dephasing are present, such as scattering from phonons or impurities, which will reduce the lifetime (or equivalently the mean free path). For the experimentally determined mean free path of $l = 81$ nm [17], the dephasing rate is $\Gamma = 0.05$ meV. One may derive alternate forms of the BCS gap equations which allow for the inclusion of non-zero damping, as given in Eqs. (3.16) and (3.17).

$$\Delta_{\mathbf{k}}^\alpha = -\frac{x_{\mathbf{k}}^2}{N} {\sum_{\mathbf{p}}}' V_{SC}(\mathbf{p}-\mathbf{k}) \left\{ -x_{\mathbf{p}}^2 \frac{\Delta_{\mathbf{p}}^\alpha}{2\Omega_{\mathbf{p}}^\alpha} \int_{-\infty}^\infty \frac{d\omega}{\pi} \mathrm{Im} G_a(\mathbf{p}, \omega) \tanh\left(\frac{\omega}{2k_B T}\right) \right.$$
$$\left. - w_{\mathbf{p}}^2 \frac{\Delta_{\mathbf{p}}^\beta}{2\Omega_{\mathbf{p}}^\beta} \int_{-\infty}^\infty \frac{d\omega}{\pi} \mathrm{Im} G_d(\mathbf{p}, \omega) \tanh\left(\frac{\omega}{2k_B T}\right) \right\} \quad (3.16)$$

$$\Delta_{\mathbf{k}}^\beta = -\frac{w_{\mathbf{k}}^2}{N} {\sum_{\mathbf{p}}}' V_{SC}(\mathbf{p}-\mathbf{k}) \left\{ -x_{\mathbf{p}}^2 \frac{\Delta_{\mathbf{p}}^\alpha}{2\Omega_{\mathbf{p}}^\alpha} \int_{-\infty}^\infty \frac{d\omega}{\pi} \mathrm{Im} G_a(\mathbf{p}, \omega) \tanh\left(\frac{\omega}{2k_B T}\right) \right.$$
$$\left. - w_{\mathbf{p}}^2 \frac{\Delta_{\mathbf{p}}^\beta}{2\Omega_{\mathbf{p}}^\beta} \int_{-\infty}^\infty \frac{d\omega}{\pi} \mathrm{Im} G_d(\mathbf{p}, \omega) \tanh\left(\frac{\omega}{2k_B T}\right) \right\} \quad (3.17)$$

with

$$G_a(\mathbf{p}, \omega) = G_b(\mathbf{p}, \omega) = \frac{1}{\omega - \Omega_{\mathbf{p}}^\alpha + i\Gamma} \quad (3.18)$$

$$G_d(\mathbf{p}, \omega) = G_g(\mathbf{p}, \omega) = \frac{1}{\omega - \Omega_{\mathbf{p}}^\beta + i\Gamma} \quad (3.19)$$

Solving these equations we find that the critical temperature is suppressed to $T_c = 2.55$ K, remarkably close to the experimental value of $T_c = 2.3$ K. We may emphasize that apart from the fixing of ω_D based on the observed gap magnitude, the critical temperature was calculated using only normal state properties of the material.

A well-known result of the BCS theory is the universal relation between the gap and the critical temperature [18] of an s-wave superconductor in the weak-coupling limit:

$$\frac{2\Delta_0}{k_B T_c} = 3.53 \tag{3.20}$$

We can use this to try to understand the strength of the coupling in our model of CeCoIn$_5$. Using the theoretical results obtained above, we find that

$$\frac{2\Delta_0}{k_B T_c} = 5.43 \tag{3.21}$$

However, a straightforward comparison of these values is not possible, for it is known that in multi-band superconductors the above ratio (with Δ_0 the maximum gap considering all the bands) can exceed the BCS value significantly, even in the weak-coupling regime [19, 20]. Taking the single band d-wave result of $\frac{2\Delta_0}{k_B T_c} = 4.3$ as a lower bound, it is seen that CeCoIn$_5$ is at most moderately coupled, so that the extension of our BCS-level model to a full strong-coupling Eliashberg theory would not be likely to introduce significant changes.

3.3 Phase-Sensitive QPI

The QPI measurements in the superconducting state discussed above in Chap. 2 allow for the determination of the gap magnitude, but not its sign. This prevents them from distinguishing between the sign-changing $d_{x^2-y^2}$-symmetry and (for instance) nodal s-wave symmetry. The method of phase-sensitive QPI (PQPI) was developed to overcome this limitation [21, 22]. The assumption is made that in the absence of an external magnetic field ($B = 0$), the Bogoliubov quasiparticle scattering is dominated by purely potential defects. In the presence of a finite field ($B \neq 0$), additional magnetic scattering channels will open up, for instance, off of polarized magnetic defects. As discussed below, the scattering from magnetic defects is sensitive to a sign change in the superconducting gap $\Delta_\mathbf{k}$ in a different way than the scattering from potential defects. The effects of potential scattering were encoded in the matrix of Eq. (2.38). Analogously, one may define for magnetic scattering the matrix

$$\hat{U} = \begin{pmatrix} M_{cc} & M_{cf} & 0 & 0 \\ M_{fc} & M_{ff} & 0 & 0 \\ 0 & 0 & -M_{cc} & -M_{cf} \\ 0 & 0 & -M_{fc} & -M_{ff} \end{pmatrix} \tag{3.22}$$

3.3 Phase-Sensitive QPI

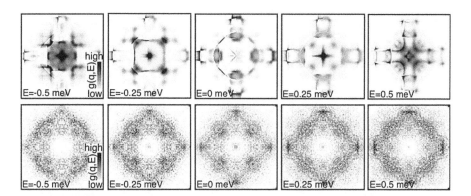

Fig. 3.5 Comparison of theoretical (top row) and experimental (bottom row) QPI spectra in a magnetic field $B = 3$ T. For the theoretical results, a low pass filter was applied to simulate the experimental resolution (right side of each panel) [16]

with M_{cc} and $M_{ff} = M_{ff}^{(0)} r_0^2$ the magnetic scattering potentials for intraband scattering, whereas $M_{fc} = M_{cf} = M_{cf}^{(0)} r_0$ describe interband scattering (recall r_0 is the expectation value of the slave boson). Here for simplicity the magnetic scatterers are assumed to split the spin up and down states, but not introduce spin-flip scattering. Thus, one has

$$\hat{g}(\mathbf{q}, E, B \neq 0) = \bar{g}_{\text{pot}}(\mathbf{q}, E) + \bar{g}_{\text{mag}}(\mathbf{q}, E, B) \quad (3.23)$$

$$\hat{g}(\mathbf{q}, E, B = 0) = \bar{g}_{\text{pot}}(\mathbf{q}, E) \quad (3.24)$$

The magnetic field dependence of $\bar{g}_{\text{mag}}(\mathbf{q}, E, B)$ is not known, but one may take $M_{cf}/M_{cc} = U_{cf}/U_{cc}$, $M_{ff}/M_{cc} = U_{ff}/U_{cc}$, and $M_{cc} \approx -1.7 U_{cc}$ to obtain reasonable agreement with the observed QPI pattern in a $B = 3$ T field, as shown in Fig. 3.5. The phase-sensitive QPI spectrum is defined by

$$\Delta g(\mathbf{q}, E, B) \equiv |\bar{g}(\mathbf{q}, E, B)| - |\bar{g}(\mathbf{q}, E, 0)| \quad (3.25)$$

Note that Eq. (2.33) for the QPI signal only includes terms which have either normal Green's functions or have anomalous Green's functions, but not combinations of the two. The crucial point to recognize is that the latter terms contain phase-sensitive information. To illustrate this, consider the contribution to the QPI signal coming from scattering within the α-band, due to the anomalous Green's functions,

$$\bar{g}_F^{\alpha\alpha}(\mathbf{q}, E) = \text{Im}\left[\frac{1}{N}\sum_{\mathbf{k}} B(\mathbf{k}, \mathbf{k}+\mathbf{q}) \frac{\Delta_{\mathbf{k}}^{\alpha}}{(E+i\Gamma)^2 - (\Omega_{\mathbf{k}}^{\alpha})^2} \frac{\Delta_{\mathbf{k}+\mathbf{q}}^{\alpha}}{(E+i\Gamma)^2 - (\Omega_{\mathbf{k}+\mathbf{q}}^{\alpha})^2}\right]$$
$$(3.26)$$

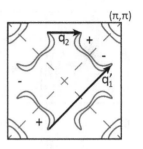

Fig. 3.6 Dominant PQPI vectors at $E = -0.5$ meV [16]

In Eq. (3.26), the factor $B(\mathbf{k}, \mathbf{k}+\mathbf{q})$ has a complicated form involving the tunneling parameters t_c, t_f, heavy fermion coherence factors $w_\mathbf{k}$, $x_\mathbf{k}$, and scattering strengths U and M. However, none of these factors are sensitive to the phase of the superconducting gaps. The remaining part of (3.26) is sensitive to the phase, thanks to the product of the two gap functions $\Delta_\mathbf{k}^\alpha \Delta_{\mathbf{k}+\mathbf{q}}^\alpha$. For instance, at an energy of $E = -0.5$ meV, the major scattering processes take place along the vectors $\mathbf{q}_{1,2}$ as shown in Fig. 3.6 (strictly speaking, the Umklapp vector $q_1' = (2\pi, 2\pi) - q_1$ is shown for convenience). Notice how q_1 connects points of different phase of the superconducting gap, whereas q_2 connects points with the same phase. Thus, the sign of the QPI contribution $\bar{g}_F^{\alpha\alpha}(\mathbf{q}_1, E)$ is different than that of $\bar{g}_F^{\alpha\alpha}(\mathbf{q}_2, E)$. On the other hand, terms involving only the normal Green's functions take the form

$$\bar{g}_G^{\alpha\alpha}(\mathbf{q}, E) = \text{Im}\left[\frac{1}{N}\sum_\mathbf{k} A(\mathbf{k},\mathbf{k}+\mathbf{q})\frac{\omega + i\Gamma + E_\mathbf{k}^\alpha}{(\omega+i\Gamma)^2 - (\Omega_\mathbf{k}^\alpha)^2}\frac{\omega + i\Gamma + E_{\mathbf{k}+\mathbf{q}}^\alpha}{(\omega+i\Gamma)^2 - (\Omega_{\mathbf{k}+\mathbf{q}}^\alpha)^2}\right] \quad (3.27)$$

which does not contain any phase information, since it depends only on $|\Delta_\mathbf{k}^{\alpha,\beta}|^2$ (inside of the energies $\Omega_\mathbf{k}^\alpha$). We next note that the form of the scattering matrices implies that the anomalous components $\bar{g}_F^{\alpha\alpha}(\mathbf{q}, E)$ have opposite signs for potential and magnetic scattering:

$$\bar{g}_{\text{pot}}(\mathbf{q}, E) = U_{cc}[\bar{g}_G(\mathbf{q}, E) - \bar{g}_F(\mathbf{q}, E)] \quad (3.28)$$

$$\bar{g}_{\text{mag}}(\mathbf{q}, E) = M_{cc}[\bar{g}_G(\mathbf{q}, E) + \bar{g}_F(\mathbf{q}, E)] \quad (3.29)$$

using the fact that $M_{cf}/M_{cc} = U_{cf}/U_{cc}$ and $M_{ff}/M_{cc} = U_{ff}/U_{cc}$. Furthermore, direct calculation shows that for $M_{cc} = -2U_{cc}$ the terms in (3.25) have opposite signs, and so

$$\begin{aligned}\Delta g(\mathbf{q}, E, B) &= \text{sgn}[\bar{g}(\mathbf{q}, E, B)]\{U_{cc}[\bar{g}_G(\mathbf{q}, E) - \bar{g}_F(\mathbf{q}, E)] \\
&\quad + M_{cc}[\bar{g}_G(\mathbf{q}, E) + \bar{g}_F(\bar{q}, E)] + U_{cc}[\bar{g}_G(\mathbf{q}, E) - \bar{g}_F(\mathbf{q}, E)]\} \\
&= \text{sgn}[\bar{g}(\mathbf{q}, E, B)][(2U_{cc} + M_{cc})\bar{g}_G(\mathbf{q}, E) + (M_{cc} - 2U_{cc})\bar{g}_F(\mathbf{q}, E)] \\
&= 2\text{sgn}[\bar{g}(\mathbf{q}, E, B)]M_{cc}\bar{g}_F(\mathbf{q}, E) \quad (3.30)\end{aligned}$$

3.4 Spin Excitations in CeCoIn$_5$

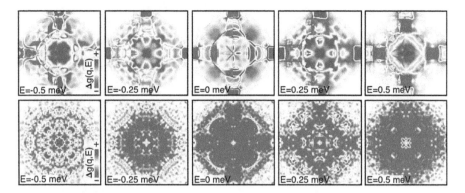

Fig. 3.7 Comparison of experimental and theoretical results for PQPI (assuming $d_{x^2-y^2}$ symmetry). For the theoretical results, a low pass filter was applied to simulate the experimental resolution (right side of each panel) [16]

Thus for this case in particular $\Delta g(\mathbf{q}, E, B)$ is proportional to $\bar{g}_F(\mathbf{q}, E)$ alone and so it is sensitive to the phase of the superconducting order parameter. For $M_{cc} = -1.7 U_{cc}$ the calculated and experimental PQPI spectra are in good agreement, as shown in Fig. 3.7. Since the ratio M_{cc}/U_{cc} used to obtain agreement with the experiment is close to the value -2, the PQPI for these parameters is indeed phase-sensitive (dominated by $\bar{g}_F(\mathbf{q}, E)$). Analyzing the dominant wavevectors $\mathbf{q}_{1,2}$ introduced above, we see that $\Delta g(\mathbf{q}_1, E, B) < 0$ whereas $\Delta g(\mathbf{q}_2, E, B) > 0$. This indicates that the gap function changes sign for scattering between $\mathbf{k}$-points on the Fermi surface connected by $\mathbf{q}_1$, while the sign remains the same for scattering of $\mathbf{q}_2$. This is shown in Fig. 3.8 for the energy $E = -0.5$ meV where these $\mathbf{q}$-vectors dominate the scattering. Thus, the superconducting gap indeed exhibits a sign-changing $d_{x^2-y^2}$ symmetry. Another method of checking the symmetry is to compare these results with the PQPI obtained from the assumption that the symmetry is nodal s-wave (i.e., the same sign, but varying magnitude, across the Fermi surface). The calculation of $\Delta g(\mathbf{q}, E, B)$ for this gap structure is given in Fig. 3.9, and is clearly inconsistent with the experimental results, strengthening the proposal that the gap changes sign along the Fermi surface. Finally, we note that the details of the experimental analysis of the PQPI can be found in Ref. [16].

3.4 Spin Excitations in CeCoIn$_5$

The study of spin excitations in the superconducting state has played an important role in both conventional and unconventional superconductors. In the former, the prediction and subsequent observation of the Hebel-Slichter peak in the spin-lattice relaxation rate below the transition temperature served to garner support for the BCS theory [23]. In unconventional superconductors, the existence of a "magnetic

Fig. 3.8 Experiment and theory for PQPI on CeCoIn$_5$, indicating the important **q**-vectors at $E = -0.5$ meV [16] (**a**) theory (**b**) experiment

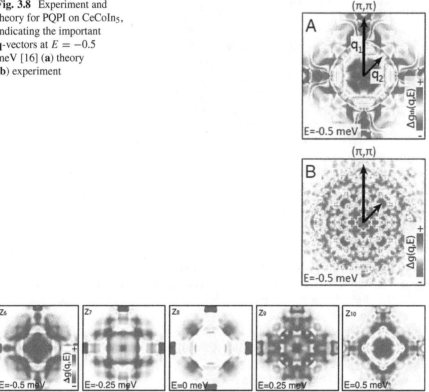

Fig. 3.9 Theoretical PQPI on CeCoIn$_5$, obtained by assuming s-wave symmetry of the superconducting gap. These results are clearly inconsistent with the experimental results in Fig. 3.7 [16]

resonance peak" in the neutron scattering data in the superconducting states of cuprates, heavy fermions, and iron pnictides has led to the speculation that the pairing mechanisms are related in each case [24]. In the following we discuss the predictions for the magnetic resonance peak and NMR relaxation rate obtained with the theoretical model developed for CeCoIn$_5$.

3.4.1 Magnetic Resonance Peak

The observation of a peak in the inelastic neutron scattering data in the superconducting state of CeCoIn$_5$ [25] has led to several proposed explanations of the phenomenon. One possibility is that the peak is a spin exciton arising in an RPA calculation of the spin susceptibility [26]. An alternative picture envisions the resonance as a magnon that becomes undamped in the superconducting state [27].

3.4 Spin Excitations in CeCoIn$_5$

These two scenarios will be explored further in Chap. 4. For now, we consider the description of the spin exciton with the model developed in Chap. 2 and the present chapter. The localized f-electrons provide the largest contribution to the magnetic susceptibility of CeCoIn$_5$, and hence we neglect the explicit contributions from the c-band or from interband terms in the following. The magnetic susceptibility $\chi(\mathbf{r} - \mathbf{r}', \tau)$ then is defined via

$$\chi(\mathbf{r} - \mathbf{r}', \tau) = \langle T_\tau \mathbf{S}_\mathbf{r}^f(\tau) \cdot \mathbf{S}_{\mathbf{r}'}^f(0) \rangle$$

$$= \frac{1}{2} \langle T_\tau S_\mathbf{r}^+(\tau) S_{\mathbf{r}'}^-(0) \rangle + \langle T_\tau S_\mathbf{r}^-(\tau) S_{\mathbf{r}'}^+(0) \rangle + \langle T_\tau S_\mathbf{r}^z(\tau) S_{\mathbf{r}'}^z(0) \rangle$$

$$= \chi^\pm(\mathbf{r} - \mathbf{r}', \tau) + \chi^\mp(\mathbf{r} - \mathbf{r}', \tau) + \chi^{zz}(\mathbf{r} - \mathbf{r}', \tau) \quad (3.31)$$

where τ is the imaginary time in the Matsubara formalism for finite temperature calculations, $\mathbf{S}^f$ are the spin operators for the f-electrons with z-components S^z and transverse components $S^\pm = S^x \pm i S^y$. For non-interacting, but hybridized, heavy quasiparticles the retarded magnetic susceptibility in the superconducting state is found to be

$$\chi_0^{SC}(\mathbf{q}, \omega) = -\frac{1}{2N} \sum_\mathbf{k} \sum_{i,j=\alpha,\beta} \zeta_{\mathbf{k},i}^2 \zeta_{\mathbf{k}+\mathbf{q},j}^2 \left\{ \left(1 + \frac{E_\mathbf{k}^i E_{\mathbf{k}+\mathbf{q}}^j + \Delta_\mathbf{k}^i \Delta_{\mathbf{k}+\mathbf{q}}^j}{\Omega_\mathbf{k}^i \Omega_{\mathbf{k}+\mathbf{q}}^j} \right) \frac{n_F(\Omega_\mathbf{k}^i) - n_F(\Omega_{\mathbf{k}+\mathbf{q}}^j)}{\omega + i\delta + \Omega_\mathbf{k}^i - \Omega_{\mathbf{k}+\mathbf{q}}^j} \right.$$

$$+ \left(1 - \frac{E_\mathbf{k}^i E_{\mathbf{k}+\mathbf{q}}^j + \Delta_\mathbf{k}^i \Delta_{\mathbf{k}+\mathbf{q}}^j}{\Omega_\mathbf{k}^i \Omega_{\mathbf{k}+\mathbf{q}}^j} \right) \frac{(\Omega_\mathbf{k}^i + \Omega_{\mathbf{k}+\mathbf{q}}^j)}{(\omega + i\delta)^2 - (\Omega_\mathbf{k}^i + \Omega_{\mathbf{k}+\mathbf{q}}^j)^2}$$

$$\left. \left[1 - n_F(\Omega_\mathbf{k}^i) - n_F(\Omega_{\mathbf{k}+\mathbf{q}}^j) \right] \right\} \quad (3.32)$$

with

$$\zeta_{\mathbf{k},i}^2 = \begin{cases} w_\mathbf{k}^2 & \text{if } i = \alpha \\ x_\mathbf{k}^2 & \text{if } i = \beta \end{cases} \quad (3.33)$$

with $\delta = 0^+$. Including the spin-flip interaction (Eq. (3.4)) between the quasiparticles and performing a standard RPA summation, one obtains the full susceptibility

$$\chi_{SC,RPA}^\pm(\mathbf{q}, \omega) = \frac{1}{2} \frac{\chi_0^{SC}(\mathbf{q}, \omega)}{1 + \bar{I}_0(\mathbf{q}) \chi_0^{SC}(\mathbf{q}, \omega)} \quad (3.34)$$

where we defined $\bar{I}_0(\mathbf{q}) \equiv I_0(\mathbf{q})/2$. Here the important thing to notice is that $\chi_{SC,RPA}^\pm$ contains the bare, unrenormalized magnetic interaction, $\bar{I}_0(\mathbf{q})$, whereas the magnetic interaction extracted from the experiment is the full one. However, at the level of the RPA approximation these can be related using

$$[\bar{I}_0(\mathbf{q})]^{-1} = [\bar{I}(\mathbf{q})]^{-1} - \text{Re}\chi_0^N(\mathbf{q}, \omega = 0) \quad (3.35)$$

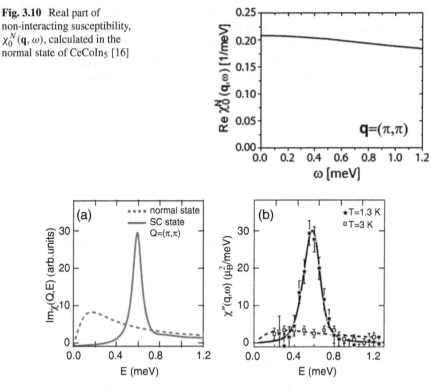

Fig. 3.10 Real part of non-interacting susceptibility, $\chi_0^N(\mathbf{q}, \omega)$, calculated in the normal state of CeCoIn$_5$ [16]

Fig. 3.11 Theoretical calculation and experimental results for the magnetic resonance peak observed in neutron scattering experiments in the superconducting state of CeCoIn$_5$ [16] (**a**) theoretical calculation (**b**) experimental results

Here the simplification of taking the normal state $\chi_0^N(\mathbf{q}, \omega)$ at $\omega = 0$ was made, which is valid since the real part is only weakly dependent on frequency for small ω, as shown in Fig. 3.10. Now computing the imaginary part of the susceptibility, which is measurable in neutron scattering experiments, using Eqs. (3.34) and (3.35), we find very good agreement between the experimental [25] and theoretical results. In particular, we note that the location of the theoretical peak at $Q = (\pi, \pi)$ in energy is very close to the experimental observation. This is shown in Fig. 3.11. The position of the resonance peak is determined by the pole of Eq. (3.34), that is, by the equation

$$1 + \bar{I}_0(\mathbf{q}) \chi_0^{SC}(\mathbf{q}, \omega) = 0 \tag{3.36}$$

In this case, since the c-electrons enter only through the non-interacting susceptibilities, and not through $\bar{I}(\mathbf{q})$, their effect on the resonance peak position is small compared to that of the f-electrons, which determine the behavior of $\bar{I}(\mathbf{q})$.

3.4.2 NMR Spin-Lattice Relaxation Rate

The spin-lattice relaxation rate measured in NMR experiments [28] can also be directly related to the spin susceptibility of Eq. (3.34).

$$\frac{1}{T_1} = \frac{k_B T}{2\hbar}(\hbar^2 \gamma_n \gamma_e)^2 \frac{A(\mathbf{q})}{N} \sum_{\mathbf{q}} \lim_{\omega \to 0} \frac{2\mathrm{Im}\chi_{SC}^{\pm}(\mathbf{q},\omega)}{\omega} \quad (3.37)$$

where γ_n and γ_e are the nuclear and electronic gyromagnetic ratios and $A(\mathbf{q})$ is the hyperfine coupling. The microscopic form of $A(\mathbf{q})$ is unknown, so that for simplicity we take a direct hyperfine coupling only, which implies momentum independence ($A(\mathbf{q}) = A_0$). For the calculation of the temperature dependence of $1/T_1$ we first determined the temperature-dependent superconducting gap using the non-linear gap equations (3.7) and (3.8) to determine $\chi_{SC,RPA}^{\pm}(\mathbf{q},\omega)$. Because of the large temperature range involved, we now calculate the bare interaction $[\bar{I}_0(\mathbf{q})]$ using the non-interacting susceptibility in the superconducting state,

$$[\bar{I}_0(\mathbf{q})]^{-1} = [\bar{I}(\mathbf{q})]^{-1} - \mathrm{Re}\chi_0^{SC}(\mathbf{q},\omega=0) \quad (3.38)$$

while the full susceptibility is still given by Eq. (3.34). The calculated temperature dependence of $1/T_1$ is given in Fig. 3.12 where it is compared with the experimental results [28]. One notices the good agreement between the theory and experiment for the relevant temperature range. Interestingly, the power-law exponent for $1/T_1$ is found theoretically to be $\alpha \approx 2.5$, which is reduced from the value $\alpha = 3$ expected for a $d_{x^2-y^2}$-wave superconductor [29]. This is understood from the fact that at the experimentally relevant temperatures, $k_B T$ exceeds the magnitude of the gap on the β-band but is smaller than that of the α-band, leading to a superposition of the power laws appropriate for the normal ($\alpha = 1$) and the superconducting states ($\alpha = 3$).

To conclude, we have demonstrated that a number of important experimental results on the superconducting gap of CeCoIn$_5$ can be reproduced under the assumption that spin fluctuations of the f-electrons are the pairing mechanism that drive superconductivity in the material. These include the symmetry of the gap, the critical temperature, the observed QPI spectra in the superconducting state (including the phase-sensitive ones), and the spin excitations of the neutron scattering and NMR experiments. This lends considerable support to the hypothesis that spin fluctuations provide the pairing in CeCoIn$_5$ and related materials, and to the theories originally proposed for heavy fermions along these lines [5, 30].

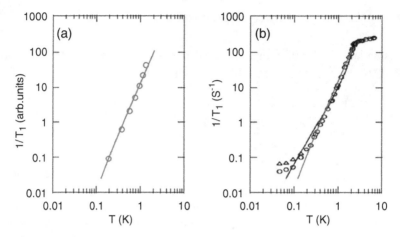

Fig. 3.12 Theoretical calculation and experimental results for the spin–lattice relaxation rate as a function of temperature in the superconducting state of CeCoIn$_5$ [16] (**a**) theoretical calculation (**b**) experimental results

References

1. F. Steglich, J. Aarts, C.D. Bredl, W. Lieke, D. Meschede, W. Franz, H. Schäfer, Superconductivity in the presence of strong Pauli paramagnetism: CeCu$_2$Si$_2$. Phys. Rev. Lett. **43**(25), 1892–1896 (1979)
2. P.G. Pagliuso, R. Movshovich, A.D. Bianchi, M. Nicklas, N.O. Moreno, J.D. Thompson, M.F. Hundley, J.L. Sarrao, Z. Fisk, Multiple phase transitions in Ce(Rh,Ir,Co)In5. Phys. B Condens. Matter *312–313*, 129–131 (2002)
3. J.L. Sarrao, J.D. Thompson, Superconductivity in cerium- and plutonium-based '115' materials. J. Phys. Soc. Jpn. **76**(5), 051013 (2007)
4. M.T. Béal-Monod, C. Bourbonnais, V.J. Emery, Possible superconductivity in nearly antiferromagnetic itinerant fermion systems. Phys. Rev. B **34**(11), 7716–7720 (1986)
5. K. Miyake, S. Schmitt-Rink, C.M. Varma, Spin-fluctuation-mediated even-parity pairing in heavy-fermion superconductors. Phys. Rev. B **34**(9), 6554–6556 (1986)
6. D.J. Scalapino, E. Loh, J.E. Hirsch, d-wave pairing near a spin-density-wave instability. Phys. Rev. B **34**(11), 8190–8192 (1986)
7. D.J. Scalapino, E. Loh, J.E. Hirsch, Fermi-surface instabilities and superconducting d-wave pairing. Phys. Rev. B **35**(13), 6694–6698 (1987)
8. P. Monthoux, G.G. Lonzarich, p-wave and d-wave superconductivity in quasi-two-dimensional metals. Phys. Rev. B **59**(22), 14598–14605 (1999)
9. M. Lavagna, A.J. Millis, P.A. Lee, d-wave superconductivity in the large-degeneracy limit of the Anderson lattice. Phys. Rev. Lett. **58**(3), 266–269 (1987)
10. P. Coleman, N. Andrei, Kondo-stabilised spin liquids and heavy fermion superconductivity. J. Phys. Condens. Matter **1**(26), 4057 (1989)
11. R. Flint, P. Coleman, Tandem pairing in heavy-fermion superconductors. Phys. Rev. Lett. **105**(24), 246404 (2010)
12. P. Coleman, A.M. Tsvelik, N. Andrei, H.Y. Kee, Co-operative Kondo effect in the two-channel Kondo lattice. Phys. Rev. B **60**(5), 3608–3628 (1999)
13. R. Flint, M. Dzero, P. Coleman, Heavy electrons and the symplectic symmetry of spin. Nat. Phys. **4**(8), 643–648 (2008)

References

14. R. Flint, A.H. Nevidomskyy, P. Coleman, Composite pairing in a mixed-valent two-channel Anderson model. Phys. Rev. B **84**(6), 064514 (2011)
15. O. Erten, R. Flint, P. Coleman, Molecular pairing and fully gapped superconductivity in Yb-doped $CeCoIn_5$. Phys. Rev. Lett. **114**(2), 027002 (2015)
16. J.S. Van Dyke, F. Massee, M.P. Allan, J.C.S. Davis, C. Petrovic, D.K. Morr, Direct evidence for a magnetic f-electron-mediated pairing mechanism of heavy-fermion superconductivity in $CeCoIn_5$. Proc. Natl. Acad. Sci. **111**(32), 11663–11667 (2014)
17. R. Movshovich, M. Jaime, J.D. Thompson, C. Petrovic, Z. Fisk, P.G. Pagliuso, J.L. Sarrao, Unconventional superconductivity in $CeIrIn_5$ and $CeCoIn_5$: specific heat and thermal conductivity studies. Phys. Rev. Lett. **86**(22), 5152–5155 (2001)
18. J.R. Schrieffer, *Theory of Superconductivity*, revised edn. (Perseus Books, Reading, 1999)
19. O.V. Dolgov, I.I. Mazin, D. Parker, A.A. Golubov, Interband superconductivity: contrasts between Bardeen-Cooper-Schrieffer and Eliashberg theories. Phys. Rev. B **79**(6), 060502 (2009)
20. S. Maiti, A.V. Chubukov, Relation between nodes and $2\Delta/T_c$ on the hole Fermi surface in iron-based superconductors. Phys. Rev. B **83**(22), 220508 (2011)
21. T. Hanaguri, Y. Kohsaka, M. Ono, M. Maltseva, P. Coleman, I. Yamada, M. Azuma, M. Takano, K. Ohishi, H. Takagi, Coherence factors in a high-T_c cuprate probed by quasi-particle scattering off vortices. Science **323**(5916), 923–926 (2009)
22. T. Hanaguri, S. Niitaka, K. Kuroki, H. Takagi, Unconventional s-wave superconductivity in Fe(Se,Te). Science **328**(5977), 474–476 (2010)
23. L.C. Hebel, C.P. Slichter, Nuclear spin relaxation in normal and superconducting aluminum. Phys. Rev. **113**(6), 1504–1519 (1959)
24. D.J. Scalapino, A common thread: the pairing interaction for unconventional superconductors. Rev. Mod. Phys. **84**(4), 1383–1417 (2012)
25. C. Stock, C. Broholm, J. Hudis, H.J. Kang, C. Petrovic, Spin resonance in the d-wave superconductor $CeCoIn_5$. Phys. Rev. Lett. **100**(8), 087001 (2008)
26. I. Eremin, G. Zwicknagl, P. Thalmeier, P. Fulde, Feedback spin resonance in superconducting $CeCu_2Si_2$ and $CeCoIn_5$. Phys. Rev. Lett. **101**(18), 187001 (2008)
27. A.V. Chubukov, L.P. Gor'kov, Spin resonance in three-dimensional superconductors: the case of $CeCoIn_5$. Phys. Rev. Lett. **101**(14), 147004 (2008)
28. Y. Kohori, Y. Yamato, Y. Iwamoto, T. Kohara, E.D. Bauer, M.B. Maple, J.L. Sarrao, NMR and NQR studies of the heavy fermion superconductors $CeTIn_5$(T=Co and Ir). Phys. Rev. B **64**(13), 134526 (2001)
29. D.J. Scalapino, The case for $d_{x^2-y^2}$ pairing in the cuprate superconductors. Phys. Rep. **250**(6), 329–365 (1995)
30. P. Monthoux, G.G. Lonzarich, Magnetically mediated superconductivity in quasi-two and three dimensions. Phys. Rev. B **63**(5), 054529 (2001)

Chapter 4
Real and Momentum Space Probes in CeCoIn$_5$: Defect States in Differential Conductance and Neutron Scattering Spin Resonance

4.1 Real-Space Study of Defects by STM

The development of scanning tunneling microscopy, specifically its spectroscopic imaging mode of operation, has enabled detailed studies of the local electronic structures of many superconductors. In particular, it is now possible to examine the detailed changes in the electronic structure in the vicinity of defects, whether point-like or extended. Defect physics has traditionally played an important role in the study of superconductivity, but its experimentally accessible effects were limited to the modification of bulk properties such as the critical temperature T_c. With STM experiments, the ability to measure precise local densities of states provides new and strong constraints on theoretical models of defects in superconductors. In particular, the local response of unconventional superconductors to defects can be a signature of the underlying superconducting gap symmetry [1]. The existence of sub-gap impurity states is one consequence of defects in superconductors which has been confirmed in a number of cases.

4.1.1 Model

To investigate the form of the differential conductance, dI/dV, in CeCoIn$_5$ in the normal and superconducting states, we start from the electronic band structure extracted from QPI spectroscopy (Chap. 2), described by the mean-field Hamiltonian $H_{tot}^{MF} = H_K^{MF} + H_{SC}^{MF}$ as given in Eqs. (2.18) and (2.19).

Zhou et al. performed STM-STS measurements on CeCoIn$_5$ in the normal state at several different temperatures [2]. In particular, they found the development of a feature at 5.3 K which they attributed to the possibility of a pseudogap regime in the material, similar to the cuprates [3]. A similar structure was also observed when superconductivity was suppressed by a magnetic field. These experimental results

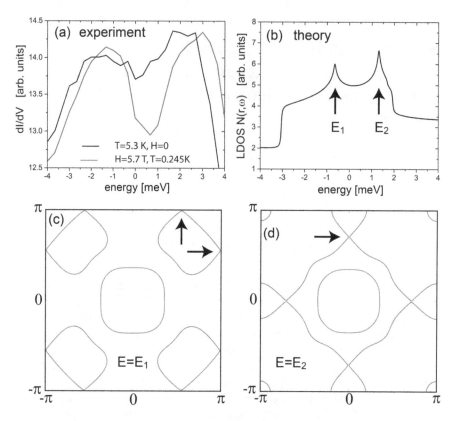

Fig. 4.1 (a) Experimental and (b) theoretical differential conductance dI/dV in the normal state of CeCoIn$_5$. (c), (d) Equal energy contours of the heavy quasiparticle bands indicating the van Hove singularities (black arrows) at energies E_1 and E_2, respectively [4]

are shown in Fig. 4.1a. To test this interpretation, we calculated the expected dI/dV in the normal state based on the electronic bandstructure extracted in Chap. 2, as shown in Fig. 4.1b. We note that a similar two-peak structure is found in the calculations as was observed experimentally. However, in this case the structure does not arise from pseudogap physics, but simply reflects the existence of van Hove singularities due to the flatness of the bands in the hybridized heavy Fermi liquid state. These singularities are clearly visible in the equal energy contours of Fig. 4.1c and d, as indicated by the arrows. Thus we propose that the signatures observed by Zhou et al. in the normal state are not due to a pseudogap, but are consequences of the hybridized band structure in this heavy fermion compound.

Next we discuss the form of dI/dV obtained in the superconducting state of CeCoIn$_5$. The Fermi surface of Chap. 2 and superconducting gap computed in Chap. 3 are presented for reference in Fig. 4.2a,b, and will be used in the subsequent discussion. Recall that there are gaps on three different sheets of the Fermi surface: two corresponding to the α-band with $\Delta_{\max}^{\alpha_1} = 0.6$ meV and $\Delta_{\max}^{\alpha_2} = 0.2$ meV,

4.1 Real-Space Study of Defects by STM 49

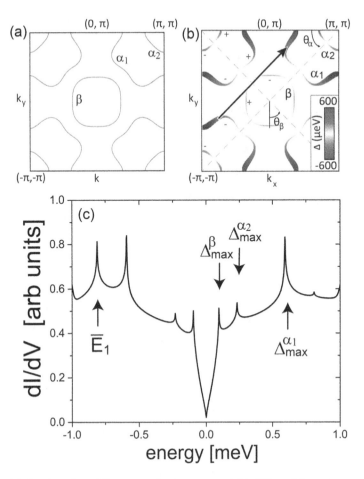

Fig. 4.2 (a) Fermi surface, (b) superconducting gap, and (c) differential conductance in the superconducting state of CeCoIn$_5$ [4]

respectively, and one from the β-band with $\Delta^{\beta}_{max} = 0.1$ meV. This leads to three sets of coherence peaks in the tunneling data, as shown in Fig. 4.2c. There is also an additional peak indicated at the energy $\bar{E}_1$, which is associated with the van Hove singularity in the normal state at $E = E_1$. One also notices a non-linear rise in the differential conductance at the lowest energies, due to the higher-harmonic form of the gap in the β-band (Eq. (3.15)).

A detailed comparison of the theoretical and experimental differential conductance is given in Fig. 4.3a, in which the calculations have been broadened by a quasiparticle damping of $\Gamma = 0.06$ meV to mimic the experimental resolution. This suppresses and smooths out the three sets of coherence peaks; in particular, those due to the α_2 and β gaps merge into a kink at $E \approx \pm 0.15$ meV, as indicated by the arrows in Fig. 4.3a. The appearance of the kink follows from the differing

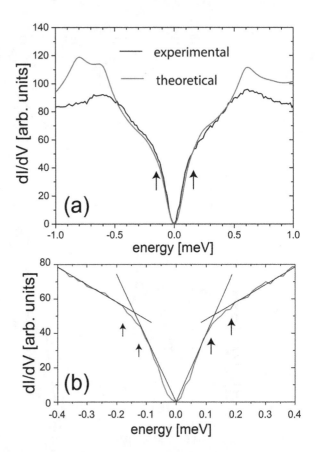

Fig. 4.3 Comparison of theoretical and experimental dI/dV in the superconducting state in the energy ranges (**a**) ± 1.0 meV and (**b**) ± 0.4 meV [4]

energy scales of the α_1 and β gaps, as indicated by the linear fits in Fig. 4.3b. For $0.2\,\text{meV} < E < 0.6\,\text{meV}$, inside the α gap but outside the β gap, the slope of dI/dV is controlled by $\Delta^{\alpha_1}_{\max}$. On the other hand at low energies, $E < 0.1$ meV, the β-band dominates the behavior, leading to a steeper slope in the conductance. The departures from linearity between $E \approx \pm 0.12$ meV and $E \approx \pm 0.18$ meV are close to the predicted coherence peaks of the two smaller gaps, which suggests that higher resolution experiments may be able to see peaks at these energies.

Having investigated the differential conductance, and thus the electronic structure, far away from strong local perturbations, we now ask how these are modified in vicinity of defects. This can be addressed in the T-matrix formalism as follows. First, define the non-interacting Green's function matrix in real space for the c and f electrons,

$$\hat{g}(\mathbf{r}, \mathbf{r}', \tau, \tau',) = -\langle T_\tau \Psi_\mathbf{r}(\tau) \Psi^\dagger_{\mathbf{r}'}(\tau') \rangle \qquad (4.1)$$

4.1 Real-Space Study of Defects by STM

In this definition the spinor $\Psi_\mathbf{r}^\dagger$ is given by $\left(c_{\mathbf{r},\uparrow}^\dagger, c_{\mathbf{r},\downarrow}, f_{\mathbf{r},\uparrow}^\dagger, f_{\mathbf{r},\downarrow}\right)$. For a single defect at position **R** which is capable of scattering either the c- or f-electrons, the dressed Matsubara Green's function is obtained from a geometric series:

$$\hat{G}(\mathbf{r},\mathbf{r}',i\omega_n) = \hat{g}(\mathbf{r},\mathbf{r}',i\omega_n) + \hat{g}(\mathbf{r},\mathbf{R},i\omega_n)\left[\hat{1} - \hat{U}\hat{g}(\mathbf{R},\mathbf{R},i\omega_n)\right]^{-1} \hat{U}\hat{g}(\mathbf{R},\mathbf{r}',i\omega_n) \tag{4.2}$$

with potential scattering matrix

$$\hat{U} = \begin{pmatrix} U_c \sigma_z & 0 \\ 0 & U_f \sigma_z \end{pmatrix}. \tag{4.3}$$

Here,

$$\hat{U} = \begin{pmatrix} U_c \sigma_z & 0 \\ 0 & U_f \sigma_z \end{pmatrix}. \tag{4.4}$$

where U_c and U_f are the potentials for scattering electrons in the c- and f-bands and σ_z is a Pauli matrix. We then analytically continue from the Matsubara to the retarded Green's function, $i\omega_n \to \omega + i\Gamma$, with the dephasing Γ determined by comparison to the experimentally determined line widths.

Placing a defect that scatters only the f-electrons at the origin, we calculate the resultant local density of states for a weak potential $U_f = -5$ meV and a vacancy of the f-electron site, modeled by letting $U_f \to -\infty$. The two cases are shown in Fig. 4.4a, b, respectively. Here the local density of states is shown for sites (1,0) and (1,1), along with that of the unperturbed system. It is readily seen that the defect induces sub-gap states at the nearest neighbor site for both scattering strengths. For stronger potentials, the state is pulled down in energy towards $E = 0$, although even in the $U_f \to -\infty$ limit the state remains at finite energies due to the particle-hole asymmetry of the bandstructure.

In addition to examining the energy dependence of the states at a fixed position, much can be learned by fixing the energy and looking at the spatial structure around the defect [5]. In Fig. 4.5a, b we show the calculated spatial structure of the density of states at $E = \mp 0.05$ meV in the presence of a weak f-electron scatterer ($U_f = -5$ meV). This is to be compared with the experimental results of Zhou et al. reproduced in panels c and d of Fig. 4.5. At positive energies, both the theory and experiment show high intensity along the directions 45° from the bond directions, as well as at the origin. For negative energies, the calculations reproduce the four lobes of high intensity at nearby sites along the bond directions, but fail to generate the suppression of the density of states at the defect site which is seen experimentally. There are a number of possible reasons for this discrepancy, such as the use of point-like rather than extended defects in the calculations (whereas the actual defects in experiment are clearly extended), or the lack of more complicated

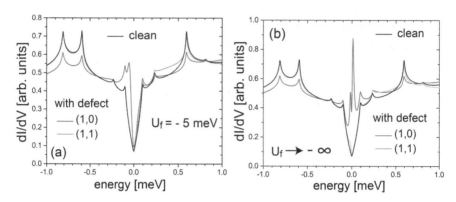

Fig. 4.4 Differential conductance in the superconducting state in the presence of a defect [4]

scattering involving the conduction electrons. Another possibility for improving the agreement would be to include the Wannier wavefunctions of the appropriate orbitals in the calculations, as was recently done with remarkable success for the cuprates [6]. Both the experimental and theoretical results shown in Fig. 4.5 agree with the expectations for $d_{x^2-y^2}$ superconductors [5]. Cuprate superconductors have also been found to agree with the $d_{x^2-y^2}$-symmetry expectations for the spatial structure of dI/dV [7].

To conclude, we demonstrated how the band structure extracted in Chap. 2 along with the superconducting gap calculated in Chap. 3 can be used to understand the real space dI/dV spectra in CeCoIn$_5$ in both the normal and superconducting states. In the former case, the pseudo-gap-like features that develop around 5.3 K can in fact be associated with the van Hove singularities of the heavy band structure. In the superconducting state, the dI/dV far from impurities is seen to carry definite signatures of the multiple gaps present in the system. Finally, the agreement between the calculated and experimental dI/dV near point-like defects reveals the presence of sub-gap impurity states, with the spatial patterns expected of $d_{x^2-y^2}$-symmetry superconductors. Taken together, these results reinforce the conclusions of the previous chapters about the band structure of CeCoIn$_5$ and the underlying magnetic f-electron pairing mechanism that produces superconductivity.

4.2 Neutron Scattering in CeCoIn$_5$

Section 3.4.1 discussed the magnetic resonance peak in the superconducting state of CeCoIn$_5$ discovered by Stock et al. [8], along with its recovery in the calculations within the model developed in Chaps. 2 and 3. Recent neutron scattering experiments on Ce$_{1-x}$Yb$_x$CoIn$_5$ by Song et al. [9] show a dispersion of this resonance, which presents an additional challenge to theorists to understand the behavior away from the commensurate antiferromagnetic wavevector Q=(0.5,0.5,0.5). The

4.2 Neutron Scattering in CeCoIn$_5$

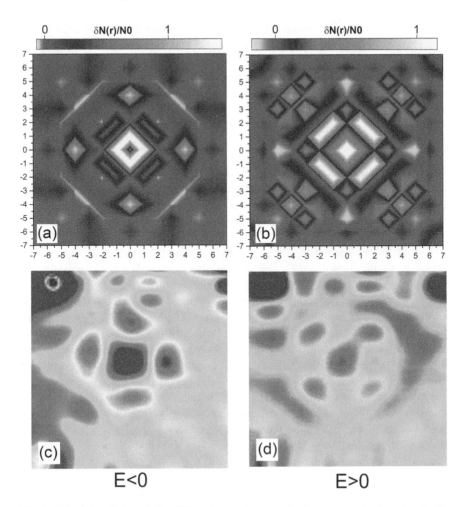

Fig. 4.5 Spatial variation of the differential conductance in the superconducting state in the presence of a defect. (**a**), (**b**) Theoretically calculated dI/dV at energies $E < 0$ and $E > 0$, respectively. (**c**), (**d**) Experimentally determined dI/dV for $E < 0$ and $E > 0$ [4]. Panels (**c**) and (**d**) are reprinted by permission from [2]

following section explores the possibility of describing the dispersing mode within the theory already developed. It is found that the most straightforward extension of the earlier work, the spin exciton scenario, fails to reproduce the correct dispersion. We then model the resonance phenomenologically as a paramagnon (damped remnant of a spin-wave from the nearby antiferromagnetic phase). Within this model we show that the observed splitting of the resonance in a magnetic field can be explained by an anisotropy of the magnetic f-electron interaction. The question of the exciton versus magnon description mirrors a discussion about a similar resonance observed in neutron scattering in the cuprates. The magnon scenario represents the strong coupling approach, in which spin-wave excitations of the f-electrons—which are damped in the normal state—become undamped in the superconducting state, leading to the resonance, as discussed below [10]. In practice, the magnon is generated by adding a term of the form Ω^2/E_F to the bare spin susceptibility. However, it was found that the weak coupling spin exciton approach automatically generates a term $\sim \Omega^2/\Delta$, which overwhelms the former term [11]. In CeCoIn$_5$ it appears the situation is reversed [12], as verified by the experiments and analysis discussed below.

Neutron scattering results from Song et al. are presented in Fig. 4.6, where the scattering wavevector is varied along the (H,H,0.5) direction in reciprocal lattice units. The experimental data are taken at a discrete set of wavelengths and energies, which are fit by the simple Gaussians shown in Fig. 4.6. Starting at the low energy side, one notices that as the energy is increased, a strong peak emerges which is centered at Q and has a maximum width at $E = 0.55$ meV. As the energy is further increased, the peak narrows slightly before splitting into two peaks dispersing away from Q. For momentum transfer fixed at Q, the scattering intensity versus energy shows a strong dependence on temperature as transition is made into the superconducting state (Fig. 4.7). This, along with the observation of similar phenomena in related unconventional superconductors [13], suggests a close tie between the resonance peak and the superconducting state. Theory developed in Sect. 3.4.1 can be immediately applied to the question of the resonance peak dispersion through the calculation of $\chi^{\pm}_{SC,RPA}(\mathbf{q},\omega)$ at wavevectors away from Q. The RPA method has been very successfully applied to other unconventional superconductors, such as the cuprates [14]. Within this approach the resonance is interpreted as a spin exciton, a collective excitation arising from the pole in the RPA susceptibility. Performing the calculation of Im$\chi^{\pm}_{SC,RPA}$ over a wide range in momentum and frequency leads to the color plots shown in Fig. 4.8a, b. It is seen that the spin exciton theory predicts a resonance with a downward dispersion. This is in marked contrast to the experimental results, overlaid with the blue line to indicate the position of the peak maximum of the resonance. Thus it appears that the simple spin exciton scenario using the dispersion and magnetic interaction extracted from the QPI experiments, along with the calculated form of the superconducting gap, is not able to reproduce the neutron scattering results away from Q.

In the absence of a microscopic theory of the spin resonance it is still possible to model the behavior phenomenologically in the spin-fermion model [10]. The idea is rooted in an analogy with the cuprate superconductors, where closeness to the parent antiferromagnetic state was proposed to engender a paramagnon resonance. CeCoIn$_5$ is also believed to be close to an antiferromagnetic state. This is evidenced by the observation of spin fluctuations in NMR and NQR measurements [15] and by the existence of non-Fermi liquid behavior in the phase diagram, which is expected in proximity to an antiferromagnetic quantum critical point [16–18]. In the normal state, the resonance is not observed, due to damping from the particle-hole continuum. However, the occurrence of superconductivity opens a gap, which allows the resonance to become undamped if its energy is below the onset of the particle-hole continuum. To study the resonance in the paramagnon framework we

4.2 Neutron Scattering in CeCoIn$_5$

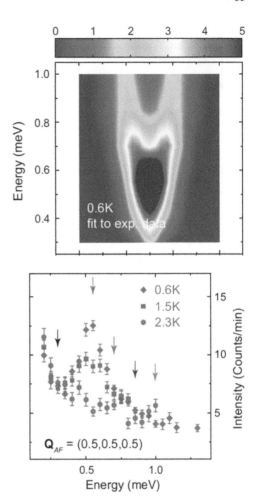

Fig. 4.6 Neutron scattering intensity as a function of wavevector (r.l.u.) and energy (meV) obtained from Gaussian fits to the experimental data

Fig. 4.7 Temperature dependence of the resonance peak in neutron scattering at the AFM wavevector Q=(π,π,π) as a function of energy

begin with the assumption that the dispersion obeys

$$\omega_{sw}^2(\mathbf{q}) = \Delta_{sw}^2 + c_{sw}^2(\mathbf{q} - \mathbf{Q}_{AF})^2 \tag{4.5}$$

with Δ_{sw} equal to the spin-wave gap and c_{sw} the corresponding velocity. The dressed spin propagator in the spin-fermion model can be written as

$$\chi^{-1} = \bar{\chi}^{-1} - \Pi \tag{4.6}$$

here $\bar{\chi}$ is the bare spin propagator and Π is the irreducible polarization operator. Re $\chi^{-1} = \bar{\chi}^{-1} -$ Re Π is determined by the fermionic excitation spectrum at all energies, and so it cannot be calculated within the low-energy model of Chaps. 2 and 3. Thus it is necessary to use a phenomenological form of the propagator,

$$\text{Re } \chi^{-1} = \bar{\chi}^{-1} - \text{Re } \Pi = \frac{\omega_{sw}^2(\mathbf{q}) - \omega^2}{\alpha} \tag{4.7}$$

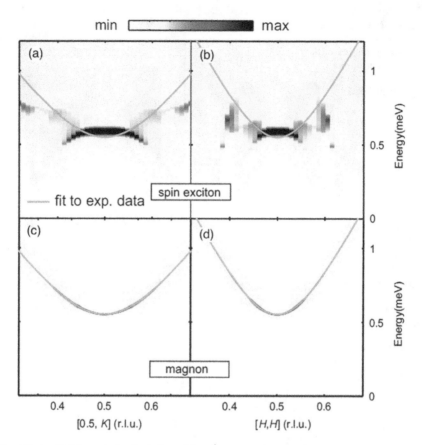

Fig. 4.8 (**a**, **b**) Theoretical calculation of $\mathrm{Im}\chi^{\pm}_{SC,RPA}$ and comparison with the experimental dispersion of the resonance peak. (**c**, **d**) Theoretical model of the resonance as a paramagnon from fits to experiment

with the dispersion given in Eq. (4.5) with the parameters $\Delta_{sw} = 0.5498$ meV and $c_{sw} = 3.2463$ Å, to reproduce the experimental results. The parameter α reflects the spectral weight of the paramagnon in the normal state. We assume in the following that the form of Eq. (4.7) is unchanged upon entry into the superconducting state. Rather, the primary effects come from Im Π, which reflects the damping of spin excitations via decay into particle-hole pairs. The lowest order expression for Π in the spin-fermion coupling g is given by

$$\Pi = g^2 \chi_0 \qquad (4.8)$$

with χ_0 the non-interacting susceptibility of Eq. (3.32). In the following we use $g^2 = 20.0$ meV^2, noting that only the width, and not the position, of the resonance is affected by this choice. The key physics that leads to the appearance of the resonance peak inside the superconducting state can be understood from Fig. 4.9.

4.2 Neutron Scattering in CeCoIn$_5$

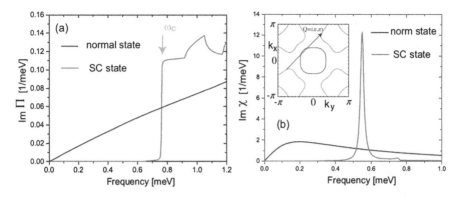

Fig. 4.9 (a) Imaginary part of the irreducible polarization Π calculated in the spin-fermion model. (b) Calculation of the resonance peak modeled as a paramagnon. Inset: wavevector **Q** of scattering processes determining Im Π and ω_c

Here, Fig. 4.9a displays Im Π in the normal and superconducting states at $\mathbf{q} = \mathbf{Q}$ as a function of energy. One notices that in the normal state the imaginary part of Π increases linearly from zero energy, i.e. damping from decay into particle-hole excitations can occur at all finite frequencies. This prevents the formation of a resonance peak at finite energy, since a spin excitation at that energy will spontaneously decay into particle-hole pairs, for $E > 0$. On the other hand, the transition to the superconducting state causes Im Π to vanish below an energy ω_c. This onset energy is determined from the fact that in the superconducting state a minimum energy is required to produce a particle-hole pair (with momenta $\mathbf{k}$ and $\mathbf{k} + \mathbf{Q}$): $\omega_c(\mathbf{Q}) = |\Delta_\mathbf{k}| + |\Delta_{\mathbf{k}+\mathbf{Q}}|$. This scattering process is shown on the Fermi surface reproduced in Fig. 4.9b. Thus, in the superconducting state the spin resonance can become undamped if its energy is below the onset ω_c. This is illustrated in the plot of Im χ reproduced in Fig. 4.9b. In particular, one notices the sudden drop in spectral weight around $E \approx 0.75$ meV, corresponding to the onset of the particle-hole continuum in Fig. 4.9a.

Moving away from the antiferromagnetic wavevector **Q** along the [1,1,0] direction, one finds that multiple onset energies appear as a consequence of the increase in the number of scattering channels connecting points on the Fermi surface separated by momentum transfer **q**. This is shown for the particular case $\mathbf{q} = 0.95\mathbf{Q}$ in Fig. 4.10a. One can clearly identify the presence of four onset energies, as indicated by the green arrows. The three high energy onsets $\omega_c^{(2)}$–$\omega_c^{(4)}$ come from scattering of $0.95\mathbf{Q}$ between different parts of the α_1 Fermi surface (Fig. 4.10c). These lead to the sudden jumps in Im Π seen in Fig. 4.10a. The jump arises from the fact the gaps at $\mathbf{k}$ and $\mathbf{k}+0.95\mathbf{Q}$ have a phase difference of π, so that the pre-factor of the second term of Eq. (3.32) does not vanish at the Fermi surface, as it would for gaps of the same sign. On the other hand, $\omega_c^{(1)}$ marks the beginning of a gradual linear onset of Im Π near $E \approx 0.4$ meV. This occurs because **q** connects momentum points on the α_2 and β Fermi surfaces, for which the signs of the gap are the same and the pre-factor of the second term vanishes on the Fermi surface.

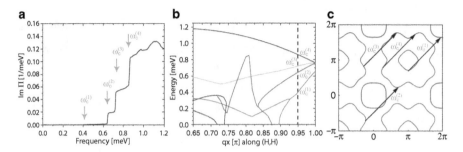

Fig. 4.10 (a) Imaginary part of Π in the superconducting state of CeCoIn$_5$ at $\mathbf{q} = 0.955\mathbf{Q}_{AF}$. Onset energies for particle-hole scattering in the superconducting state are indicated by the green arrows. (b) Momentum dependence of the onset energies in ImΠ. Dashed vertical line corresponds to the case shown in (a). (c) Scattering vectors corresponding to the onset energies in (a) shown on the Fermi surface

Fig. 4.11 The dispersion of the resonance modeled as a paramagnon, including the energies of some onsets of the particle-hole continuum. Whenever an onset is crossed there is a corresponding reduction in the amplitude of the mode

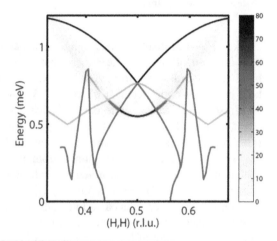

The evolution of the onset energies as a function of momentum transfer is shown in Fig. 4.10b, with the vertical dashed line indicating the case of $\mathbf{q} = 0.95\mathbf{Q}$. One sees that the original onset energy near $E \approx 0.75$ meV for $\mathbf{q} = \mathbf{Q}$ splits into three separate onset energies away from this point, and that other previously unrealized onsets emerge as new parts of the Fermi surface can be connected with a given momentum vector. The effect of these onsets on the intensity of the paramagnon resonance is demonstrated in Fig. 4.11. One notices that whenever the dispersion crosses an onset energy for the particle-hole continuum at a given wavevector, there is a subsequent loss of amplitude in the resonance mode, due to the damping produced by Im$\chi^{\pm}_{SC,RPA}$.

4.2 Neutron Scattering in CeCoIn$_5$

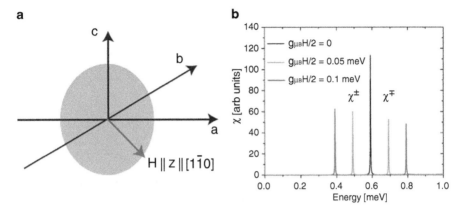

Fig. 4.12 (a) Schematic of magnetic easy plane and applied magnetic field. (b) Splitting of the resonance mode due to applied field in the presence of easy-plane magnetic anisotropy

4.2.1 Magnetic Anisotropy and External Magnetic Field

Neutron scattering experiments by Stock et al. [19] discovered that the resonance peak splits into two peaks when a magnetic field is applied in the $[1,\bar{1},0]$ direction. The splitting into two modes for the field in the ab-plane is unexpected, as the nominally spin 1 exciton should split into three peaks upon application of the external field, for a system with Heisenberg spin symmetry. However, the observation of two modes can be explained if the system possesses a magnetic easy plane perpendicular to the applied field (in this case the plane spanned by $[1,1,0]$ and $[0,0,1]$). Therefore we replace the previous magnetic interaction Hamiltonian, Eq. (3.1), with one including anisotropy and coupling to the external field

$$H_H = \sum_{\mathbf{r},\mathbf{r}'} I_{\mathbf{r},\mathbf{r}'} \mathbf{S}_\mathbf{r} \cdot \mathbf{S}_{\mathbf{r}'} + A \sum_\mathbf{r} (S^z_\mathbf{r})^2 - g\mu_B H \sum_\mathbf{r} S^z_\mathbf{r} \quad (4.9)$$

Here the choice $A > 0$ yields a hard magnetic axis along $[1,\bar{1},0]$ and an easy plane perpendicular to it (Fig. 4.12a). For convenience $[1,\bar{1},0]$ is defined as the z-direction in spin space.

With the Abrikosov pseudofermion representation also used in Chap. 2,

$$\mathbf{S}_\mathbf{r} = \frac{1}{2} \sum_{\alpha,\beta} f^\dagger_{\mathbf{r},\alpha} \sigma_{\alpha\beta} f_{\mathbf{r},\beta} \quad (4.10)$$

we may re-write Eq. (4.9) as

$$H = \frac{1}{4N} \sum_{\mathbf{k},\mathbf{l},\mathbf{q}} \left\{ I_{zz}(\mathbf{q}) \left(f^\dagger_{\mathbf{k}+\mathbf{q}\uparrow} f_{\mathbf{k}\uparrow} - f^\dagger_{\mathbf{k}+\mathbf{q}\downarrow} f_{\mathbf{k}\downarrow} \right) \left(f^\dagger_{\mathbf{l}-\mathbf{q}\uparrow} f_{\mathbf{k}\mathbf{l}\uparrow} - f^\dagger_{\mathbf{l}-\mathbf{q}\downarrow} f_{\mathbf{l}\downarrow} \right) \right.$$

$$+ I_{\pm}(\mathbf{q}) \left(f_{\mathbf{k}+\mathbf{q}\uparrow}^{\dagger} f_{\mathbf{k}\downarrow} f_{\mathbf{l}-\mathbf{q}\downarrow}^{\dagger} f_{\mathbf{l}\uparrow} + f_{\mathbf{k}+\mathbf{q}\downarrow}^{\dagger} f_{\mathbf{k}\uparrow} f_{\mathbf{l}-\mathbf{q}\uparrow}^{\dagger} f_{\mathbf{l}\downarrow} \right)$$

$$- g\mu_B H \sum_{\mathbf{k}} \left(f_{\mathbf{k}\uparrow}^{\dagger} f_{\mathbf{k}\uparrow} - f_{\mathbf{k}\downarrow}^{\dagger} f_{\mathbf{k}\downarrow} \right) \Big\} \tag{4.11}$$

where we have defined

$$I_{zz}(\mathbf{q}) = I_\mathbf{q} + A \tag{4.12}$$

$$I_{\pm}(\mathbf{q}) = I_\mathbf{q} \tag{4.13}$$

near the commensurate antiferromagnetic wavevector $\mathbf{Q}$, one has $I(\mathbf{Q}) < 0$, so that

$$|I_{zz}(\mathbf{Q})| < |I_{\pm}(\mathbf{Q})| \tag{4.14}$$

Calculating the transverse susceptibility in the RPA approximation yields

$$\chi^{\gamma}(\mathbf{q},\omega) = \frac{\chi_0^{\gamma}(\mathbf{q},\omega)}{1 + I_{\pm}\chi_0^{\gamma}(\mathbf{q},\omega)} \tag{4.15}$$

where $\gamma = \pm, \mp$ and the non-interacting transverse susceptibility is given by [20]

$$\chi_0^{\pm,\mp}(\mathbf{q},\omega) = -\frac{1}{N} \sum_{\mathbf{k}} \sum_{i,j=\alpha,\beta} \Bigg\{ C_{ij}^{+} \frac{f_{\mathbf{k}+\mathbf{q}}^{i,\pm} - f_{\mathbf{k}}^{j,\mp}}{\omega + i\delta + \xi_{\mathbf{k}+\mathbf{q}}^{i,\pm} - \xi_{\mathbf{k}}^{j,\pm}}$$

$$+ \frac{C_{ij}^{-}}{2} \frac{1 - f_{\mathbf{k}+\mathbf{q}}^{i,\mp} - f_{\mathbf{k}}^{j,\mp}}{\omega + i\delta - \xi_{\mathbf{k}+\mathbf{q}}^{i,\mp} - \xi_{\mathbf{k}}^{j,\mp}} - \frac{C_{ij}^{-}}{2} \frac{1 - f_{\mathbf{k}+\mathbf{q}}^{i,\pm} - f_{\mathbf{k}}^{j,\pm}}{\omega + i\delta + \xi_{\mathbf{k}+\mathbf{q}}^{i,\pm} + \xi_{\mathbf{k}}^{j,\pm}} \Bigg\} \tag{4.16}$$

where $\xi_{\mathbf{k}}^{i,\pm} = \Omega_{\mathbf{k}}^{i} \pm H$, $f_{\mathbf{k}}^{i,\pm} = n_F(\xi_{\mathbf{k}}^{i,\pm})$, and

$$C_{ij}^{\pm} = \frac{1}{2} \left(1 \pm \frac{E_{\mathbf{k}+\mathbf{q}}^{i} E_{\mathbf{k}}^{j} + \Delta_{\mathbf{k}+\mathbf{q}}^{i} \Delta_{\mathbf{k}}^{j}}{\Omega_{\mathbf{k}+\mathbf{q}}^{i} \Omega_{\mathbf{k}}^{j}} \right) \tag{4.17}$$

Note that $\chi_0^{\pm}$ is given by the right-hand side with the upper signs, whereas $\chi_0^{\mp}$ is given by the lower signs. Thus, the magnetic field is treated as a Zeeman splitting of the two spin directions. We can now see how magnetic anisotropy can lead to the observed two-peak structure in applied field. For large enough A, one has $I_{zz}(\mathbf{q}) = I_\mathbf{q} + A > 0$, and the longitudinal mode can be located above the onset energy of the particle-hole continuum, $\omega_c(\mathbf{Q})$. Thus, it will be strongly damped and not observable in the neutron scattering. For $H = 0$, the two transverse modes $\chi_0^{\pm,\mp}$ will be degenerate and hence produce a single peak in the spectrum, as seen in the

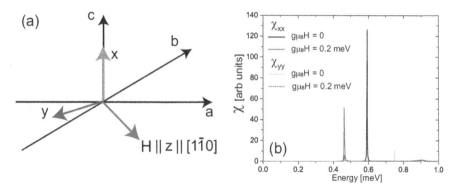

Fig. 4.13 (a) Schematic of magnetic easy-axis and applied magnetic field. (b) Splitting of the resonance mode due to applied field in the presence of easy-axis magnetic anisotropy

original experiments [8]. The applied field will then split the peak into two with energy separation increasing linearly with H, as seen in the calculation shown in Fig. 4.12b, and also observed experimentally [19].

One may contrast the behavior in the presence of a magnetic easy-plane with that in the case of an easy-axis. Assuming that the magnetic field is still along $[1,\bar{1},0]$, let the easy-axis be the magnetic x-axis (crystallographic c-axis), shown in Fig. 4.13. Then the second term in (4.9) is replaced by $A \sum_{\mathbf{r}}(S_{\mathbf{r}}^x)^2$ where now $A < 0$.

In this case the transverse susceptibilities can be written as

$$\chi_{xx}(\mathbf{q}, \omega) = \frac{1}{4} \frac{\chi_0^{\pm} + \chi_0^{\mp} + 2\chi_0^{\pm}\chi_0^{\mp}(I_{\mathbf{q}}^+ - I_{\mathbf{q}}^-)}{(1 + I_{\mathbf{q}}^+\chi_0^{\pm})(1 + I_{\mathbf{q}}^+\chi_0^{\mp}) - \chi_0^{\pm}\chi_0^{\mp}(I_{\mathbf{q}}^-)^2} \qquad (4.18)$$

$$\chi_{yy}(\mathbf{q}, \omega) = \frac{1}{4} \frac{\chi_0^{\pm} + \chi_0^{\mp} + 2\chi_0^{\pm}\chi_0^{\mp}(I_{\mathbf{q}}^+ + I_{\mathbf{q}}^-)}{(1 + I_{\mathbf{q}}^+\chi_0^{\pm})(1 + I_{\mathbf{q}}^+\chi_0^{\mp}) - \chi_0^{\pm}\chi_0^{\mp}(I_{\mathbf{q}}^-)^2} \qquad (4.19)$$

and here $I_{\mathbf{q}}^{\pm} = (I_{\mathbf{q}}^x \pm I_{\mathbf{q}}^y)/2$, $I_{\mathbf{q}}^x = I_{\mathbf{q}} + A$, and $I_{\mathbf{q}}^y = I_{\mathbf{q}}^z = I_{\mathbf{q}}$. Since $I_{\mathbf{q}} < 0$ for $\mathbf{q} \approx \mathbf{Q}$, we have $|I_{\mathbf{q}}^x| > |I_{\mathbf{q}}^{y,z}|$ and the resonance peak occurs at a lower energy for χ_{xx} than for $\chi_{yy} = \chi_{zz}$ (in zero field). As a qualitative demonstration of the behavior with an easy-axis, we set $A = -0.3$ meV and $I_{\mathbf{q}}^x$ such that the resonance in χ_{xx} occurs at $\omega = 0.6$ meV. One sees in Fig. 4.13b that for $H = 0$, χ_{xx} has the expected behavior, whereas $\chi_{yy} = \chi_{zz}$ possesses a small peak near the edge of the particle-hole continuum. Application of a finite field $H = 0.2$ meV then pushes the χ_{xx} resonance to lower energy, leaves χ_{zz} unaffected (not shown), and pushes χ_{yy} up into the particle-hole continuum where it is damped away. Thus, in this scenario there is no splitting of the resonance peak by a field applied in the $[1,\bar{1},0]$ direction, in clear contradiction with the experiments.

To conclude, in this chapter we investigated further developments of the model of CeCoIn$_5$ introduced in Chaps. 2 and 3. It was found that the model was able to quantitatively account for the real space differential conductance in the normal and superconducting states, both on clean parts of the surface and in the immediate vicinity of defects. This lent further support to the proposed low-energy band structure and microscopic pairing mechanism. We then turned to recent neutron scattering experiments on $Ce_{1-x}Yb_xCoIn_5$ and investigated the dispersion of the magnetic resonance peak in the superconducting state. It was found that the most straightforward extension of the model of Chaps. 2 and 3, the spin exciton scenario, was not able to account for the dispersion. We then modeled the resonance as a paramagnon and showed how its appearance can be understood through the opening of a gap in the particle-hole continuum below T_c. Finally, we addressed how the unexpected observation of the resonance splitting into two peaks in a magnetic field applied in the $[1,\bar{1},0]$ direction can be explained as a consequence of magnetic easy-plane anisotropy.

References

1. H. Alloul, J. Bobroff, M. Gabay, P.J. Hirschfeld, Defects in correlated metals and superconductors. Rev. Mod. Phys. **81**(1), 45–108 (2009)
2. B.B. Zhou, S. Misra, E.H. da Silva Neto, P. Aynajian, R.E. Baumbach, J.D. Thompson, E.D. Bauer, A. Yazdani, Visualizing nodal heavy fermion superconductivity in $CeCoIn_5$. Nat. Phys. **9**(8), 474–479 (2013)
3. H. Ding, T. Yokoya, J.C. Campuzano, T. Takahashi, M. Randeria, M.R. Norman, T. Mochiku, K. Kadowaki, J. Giapintzakis, Spectroscopic evidence for a pseudogap in the normal state of underdoped high-T_c superconductors. Nature **382**(6586), 51–54 (1996)
4. J.S. Van Dyke, J.C.S. Davis, D.K. Morr, Differential conductance and defect states in the heavy-fermion superconductor $CeCoIn_5$. Phys. Rev. B **93**(4), 041107 (2016)
5. S. Haas, K. Maki, Quasiparticle bound states around impurities in $d_{x^2-y^2}$-wave superconductors. Phys. Rev. Lett. **85**(10), 2172–2175 (2000)
6. A. Kreisel, P. Choubey, T. Berlijn, W. Ku, B.M. Andersen, P.J. Hirschfeld, Interpretation of scanning tunneling quasiparticle interference and impurity states in cuprates. Phys. Rev. Lett. **114**(21), 217002 (2015)
7. E.W. Hudson, K.M. Lang, V. Madhavan, S.H. Pan, H. Eisaki, S. Uchida, J.C. Davis, Interplay of magnetism and high-Tc superconductivity at individual Ni impurity atoms in $Bi_2Sr_2CaCu_2O_{8+\delta}$. Nature **411**(6840), 920–924 (2001)
8. C. Stock, C. Broholm, J. Hudis, H.J. Kang, C. Petrovic, Spin resonance in the d-wave superconductor $CeCoIn_5$. Phys. Rev. Lett. **100**(8), 087001 (2008)
9. Y. Song, J.V. Dyke, I.K. Lum, B.D. White, S. Jang, D. Yazici, L. Shu, A. Schneidewind, P. Čermák, Y. Qiu, M.B. Maple, D.K. Morr, P. Dai, Robust upward dispersion of the neutron spin resonance in the heavy fermion superconductor $Ce_{1-x}Yb_xCoIn_5$. Nat. Commun. **7**, 12774 (2016)
10. D.K. Morr, D. Pines, The resonance peak in cuprate superconductors. Phys. Rev. Lett. **81**(5), 1086–1089 (1998)
11. A. Abanov, A.V. Chubukov, A relation between the resonance neutron peak and ARPES data in cuprates. Phys. Rev. Lett. **83**(8), 1652–1655 (1999)
12. A.V. Chubukov, L.P. Gor'kov, Spin resonance in three-dimensional superconductors: the case of $CeCoIn_5$. Phys. Rev. Lett. **101**(14), 147004 (2008)
13. D.J. Scalapino, A common thread: the pairing interaction for unconventional superconductors. Rev. Mod. Phys. **84**(4), 1383–1417 (2012)

References

14. A. Abanov, A.V. Chubukov, M. Eschrig, M.R. Norman, J. Schmalian, Neutron resonance in the cuprates and its effect on fermionic excitations. Phys. Rev. Lett. **89**(17), 177002 (2002)
15. Y. Kohori, Y. Yamato, Y. Iwamoto, T. Kohara, E.D. Bauer, M.B. Maple, J.L. Sarrao, NMR and NQR studies of the heavy fermion superconductors $CeTIn_5$(T=Co and Ir). Phys. Rev. B **64**(13), 134526 (2001)
16. V.A. Sidorov, M. Nicklas, P.G. Pagliuso, J.L. Sarrao, Y. Bang, A.V. Balatsky, J.D. Thompson, Superconductivity and quantum criticality in $CeCoIn_5$. Phys. Rev. Lett. **89**(15), 157004 (2002)
17. C. Petrovic, P.G. Pagliuso, M.F. Hundley, R. Movshovich, J.L. Sarrao, J.D. Thompson, Z. Fisk, P. Monthoux, Heavy-fermion superconductivity in $CeCoIn_5$ at 2.3 K. J. Phys. Condens. Matter **13**(17), L337 (2001)
18. C.F. Miclea, M. Nicklas, D. Parker, K. Maki, J.L. Sarrao, J.D. Thompson, G. Sparn, F. Steglich, Pressure dependence of the Fulde-Ferrell-Larkin-Ovchinnikov state in $CeCoIn_5$. Phys. Rev. Lett. **96**(11), 1–4 (2006)
19. C. Stock, C. Broholm, Y. Zhao, F. Demmel, H.J. Kang, K.C. Rule, C. Petrovic, Magnetic field splitting of the spin resonance in $CeCoIn_5$. Phys. Rev. Lett. **109**(16), 167207 (2012)
20. J.-P. Ismer, I. Eremin, E. Rossi, D.K. Morr, Magnetic resonance in the spin excitation spectrum of electron-doped cuprate superconductors. Phys. Rev. Lett. **99**(4), 047005 (2007)

Chapter 5
Transport in Nanoscale Kondo Lattices

The previous chapters have thoroughly studied an archetypal heavy fermion material, $CeCoIn_5$, in the normal and superconducting states. The response to several different probes, primarily scanning tunneling spectroscopy and neutron scattering was examined and modeled within a mean-field slave boson large-N theory. A common feature of these experimental techniques is that they study the properties of the system in equilibrium. Indeed, much of the work on strongly correlated systems has focused on equilibrium behavior. The reason for this is two-fold. First, the theory of equilibrium statistical mechanics is much further developed than nonequilibrium theory, which makes it easier to calculate observables in this framework. In cases where nonequilibrium results are desired, such as the response to a time-dependent external field, the traditional approach has been to use linear response theory and the fluctuation-dissipation theorem, which allows one to obtain the first-order response by calculating only equilibrium quantities [1]. Second, because strongly correlated systems have proven difficult to understand even in the equilibrium case, it has perhaps been thought that one should not attempt a harder problem before the easier one is solved sufficiently. To this one may reply that nonequilibrium experiments present fundamentally new phenomena that can help further constrain theoretical models and lead to a more comprehensive understanding of correlated electron systems. Furthermore, while nonequilibrium calculations are generally more challenging than equilibrium ones, the rapid growth of computing power has made them less prohibitive. It seems appropriate therefore to investigate correlated systems out of equilibrium, in order to further the development of the field.

Picking up the thread of the earlier chapters, we may consider a heavy fermion system connected to metallic leads with an applied voltage bias. This will produce a charge current through the system, which can be calculated in a spatially resolved way [2]. Given the early stage of development, we do not attempt to utilize a quantitatively accurate band structure (as was done for $CeCoIn_5$ in the preceding chapters), but instead use simplified Kondo lattice model to understand general features.

Recently, the Keldysh Green function approach was applied in real space to model nanoscale simple metallic systems [2]. Even for systems free of atomic disorder, a wide variety of current flow patterns are obtained that depend on the detailed spatial structures of the electronic wavefunctions. Here we extend such calculations to heavy fermion systems at the nanoscale, which are computationally tractable relative to macroscopic systems. It is found that the presence of correlations between the conduction and localized f-electrons, encoded in the hybridization s, has a profound effect on the current patterns through a Kondo lattice, even when the currents are constrained to flow through the c-electron subsystem.

In the following we study a nanoscale heavy fermion system consisting of a square lattice of conduction electron sites coupled to a lattice of f-electrons of the same size. This system is then connected to two metallic leads, each with a constant density of states. Working in the large-N slave boson mean field theory as in earlier chapters, the Hamiltonian is

$$H = -\mu \sum_i c_i^\dagger c_i - t \sum_{<i,j>} c_i^\dagger c_j + \sum_i s_i f_i^\dagger c_i + \sum_i \varepsilon_i f_i^\dagger f_i$$
$$+ H_{lead} - t_l \sum_i c_i^\dagger d_i + h.c. \tag{5.1}$$

where c_i, f_i ($c_i^\dagger$, $f_i^\dagger$) annihilate (create) conduction and f-electrons at site i in the system, respectively, and d_i ($d_i^\dagger$) annihilates (creates) an electron in the lead at the site connected to the c-electron site i of the system. The current is calculated in the non-equilibrium Keldysh Green's function formalism in real space [3, 4] according to the expression

$$I_{rr'} = -2\frac{e}{\hbar} t \int_{-\infty}^{\infty} \frac{d\omega}{2\pi} \mathrm{Re}\left[\hat{G}_{rr'}^<(\omega)\right] \tag{5.2}$$

where $\hat{G}^<$ is the full lesser Green's function matrix including the leads and the heavy fermion system. This formalism is explained in more detail in Appendix A.

5.1 Transport in a Clean System

We start with a clean (defect-free) system with unhybridized and hybridized Fermi surfaces for an infinitely large system shown in Fig. 5.1. The hybridization s_i and f-electron chemical potential ε_i are determined from the self-consistent equations of the equilibrium mean field theory [5]. For a finite size system, the lack of periodic boundary conditions causes s_i and ε_i to vary spatially, even in the absence of defects (Fig. 5.2). Notably, the hybridization is zero along the edge of the system. With narrow leads (width of one site) attached to the middle

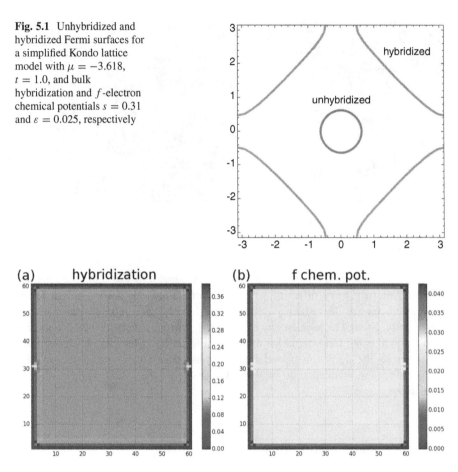

Fig. 5.1 Unhybridized and hybridized Fermi surfaces for a simplified Kondo lattice model with $\mu = -3.618$, $t = 1.0$, and bulk hybridization and f-electron chemical potentials $s = 0.31$ and $\varepsilon = 0.025$, respectively

Fig. 5.2 Self-consistently determined spatial variation of the mean-field (**a**) hybridization, s, and (**b**) f-electron chemical potential, ε_f, for a 61×61 Kondo lattice

of the left and right edges of a 61×61 system, we calculate the resulting spatially-resolved current pattern. This is presented in Fig. 5.3. For the system with correlations, that is, non-zero hybridization between c-electron and f-electron states (Fig. 5.3a), the current pattern displays a characteristic diamond shape. This is due to the velocity of the hybridized heavy quasiparticle states. At low temperatures, the transport will be dominated by the low energy excitations of electrons near the hybridized Fermi surface of Fig. 5.1. Since the electron velocity is given by $\mathbf{v}(\mathbf{k}) = (1/\hbar)\nabla_{\mathbf{k}} E(\mathbf{k})$, one sees that the typical velocities of the quasiparticles obey $|k_x| \approx |k_y|$, explaining the diagonal trajectories of the currents in Fig. 5.3a. At temperatures above the coherence temperature, T_{coh}, of the Kondo lattice, the magnetic moments are not screened by the conduction electrons. We model this situation by setting the hybridization $s = 0$, i.e. decoupling the c- and f-electron subsystems completely. In this case, the

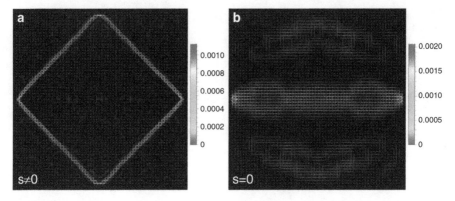

Fig. 5.3 Charge currents through a clean Kondo lattice ($N_x = 61$) attached to narrow leads for (**a**) $s_i \neq 0$ and (**b**) $s_i = 0$. Lead-system coupling is $t_l = 1.0t$

c-electron system behaves as a simple metal. As shown in Fig. 5.3b, the resulting current pattern is very different when the correlations are absent. In particular, the current path is less sharply defined, and largely goes through the center of the system. Though experimentally challenging, current patterns have been imaged using SQUIDs [6] (an alternative method using an STM has also been proposed [7]), and so the transition between the correlated Kondo lattice state and high temperature uncorrelated state may be observable in the modification of current flow in the system.

5.2 Transport with Defects

It is natural to ask how the current flow is modified by defects that are invariably present in real systems. In particular, the effect of non-zero hybridization on the response of the system to defects provides interesting signatures of the correlations in the system. To consider this, we introduce an f-electron vacancy (a Kondo hole) directly in the path of the current through the lower branch, at the site indicated by the purple star in Fig. 5.4. The resulting current pattern is presented in Fig. 5.4, which shows the dramatic modification caused by the vacancy (Fig. 5.4b) compared to the clean case (Fig. 5.4a). In interpreting these results it is important to note that because the f-electrons lack an inter-site hopping term in Eq. (5.1), the current is forced to flow entirely through the c-electron subsystem. The only effect of the f-electron vacancy is to locally modify the hybridization, but nevertheless this is found to induce strong changes in the resulting current pattern. On the other hand, the total current through the system remains almost the same, changing from $0.0011t$ in the clean case to $0.00109t$ with the defect.

To understand the qualitative differences that arise by varying the width of the attached leads, in the following we make the leads the same width as the

5.2 Transport with Defects

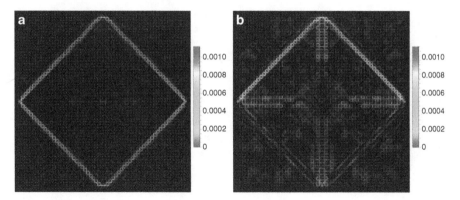

Fig. 5.4 Comparison of the currents in a hybridized Kondo lattice ($s_i \neq 0$) for (**a**) the clean system (**b**) a system with a defect at the site indicated by the purple star

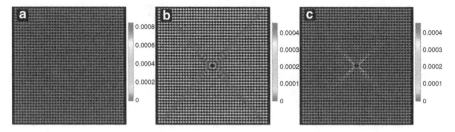

Fig. 5.5 Currents in a 41 × 41 site system with wide leads and periodic boundary conditions in the y-direction. Current patterns are shown for (**a**) a clean system, (**b**) a Kondo hole at the center of the system ($s = 0$), and (**c**) a Kondo hole with phonon coupling $\gamma = 10^{-5}t$

system itself. We keep the leads attached to the left and right edges of the system. Furthermore, we apply periodic boundary conditions in the y direction, since we are less interested in the modification of the current near the edge of the sample than we are in the change in current flow around a defect. In the absence of a defect, the current flows uniformly across the sample, as expected from symmetry (Fig. 5.5a). By introducing an f-electron vacancy at the center of the system, we obtain the current patterns shown in Fig. 5.5b,c. In these panels, the defect is modeled as a Kondo hole, by setting $s = 0$ at the central site. In Fig. 5.5b one notices that the defect induces changes in the current pattern out to the edge of the system. In Fig. 5.5c we coupled the system to a set of local phonon modes at each site, using the high temperature approximation developed in Ref. [8] (see Appendix A). The scattering of the electrons by phonons introduces a finite mean free path, randomizing the phase of the electronic wavefunctions over distances greater than this length. This suppresses the long-range effects of coherent scattering off of the defect, as seen in Fig. 5.5c.

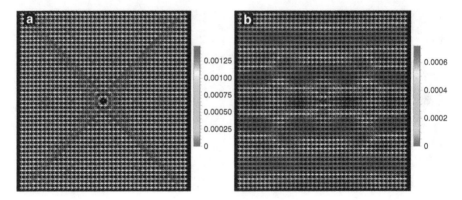

Fig. 5.6 Currents for a system with wide leads and a c-electron scatterer ($u_0 = 5.0t$) with (**a**) $s_i \neq 0$ and (**b**) $s_i = 0$

The behavior of a system with defects may also be sensitive to the presence of correlations. To study this, we consider replacing an f-electron site with a non-magnetic c-electron scatterer. In the Kondo-screened state, there is a localized c-electron potential in addition to there being zero hybridization at the site of the defect. While the changes due to the defect still extend to the edge of the system (Fig. 5.6a), the current patterns are considerably more uniform than in the state where correlations are absent ($s = 0$), shown in Fig. 5.6b. In the latter case the hybridization is everywhere zero, but the translational symmetry of the current pattern is still broken due to the c-electron scattering potential. The wavelength of the current oscillations in the vertical direction is approximately 10 lattice spacings, corresponding to the de Broglie wavelength of the c-electrons at the Fermi surface. An experimental determination of the current pattern as a function of temperature may therefore be able to sense the development of a coherent Kondo lattice.

5.3 Multiple Defects

The real-space approach adopted here easily allows for the examination of systems with multiple defects. For concreteness, consider the current through a system attached to wide leads with 1% of the f-electron sites replaced by non-magnetic conduction electron scatterers. Figure 5.7 shows the result of calculations for this case, which reveal complicated current patterns in both the correlated and uncorrelated states. The purple circles indicate the locations of the defects. In Fig. 5.7b, the correlated state, one notices the presence of gaps in the current pattern, in addition to various "hot spots" where the current is large. In contrast, the system without correlations in Fig. 5.7a does not show changes quite as drastic, except for a nearly complete suppression of the current flowing through the defect sites.

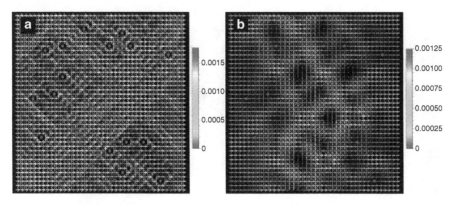

Fig. 5.7 Currents for a defect system, with wide leads and multiple defects in the (**a**) correlated and (**b**) uncorrelated states

5.4 Hopping Within the f-Band

Up until this point we have restricted our consideration to systems in which a completely localized f-electron level hybridizes with the conduction band. However, it is also conceivable that the f-band could have a narrow dispersion allowing for direct hopping between f-electron sites. We introduce an additional term $-t_f \sum_{<i,j>} f_i^\dagger f_j$ into the Hamiltonian equation (5.1) to couple the neighboring f-electrons, but keep the leads coupled to the c-electrons only. Figure 5.8a,b shows the currents flowing in both the c- and f-electron subsystems, respectively. One notices that the current pattern is sharper and its magnitude larger in the f-electron subsystem than for the c-electrons (after entering the system through the left lead attached to the c-electron site, the majority of the current immediately flows into the f-system through the hybridization at that site).

5.5 Self-Consistency with Finite Bias

In the above results, the local hybridizations and f-electron chemical potentials were determined self-consistently in equilibrium. These parameters were then fixed while the bias was applied and the current patterns calculated. While this procedure may be justified in the limit of small biases, in general the hybridization and chemical potential will change in the non-equilibrium state. To study this effect, we have also performed fully self-consistent calculations on a smaller system, using narrow leads. We first discuss the self-consistency equations out of equilibrium and then the numerical results in this case. In equilibrium it suffices to consider the imaginary time Matsubara Green's functions to develop a self-consistent large-N mean-field theory at finite temperatures. These are then analytically continued

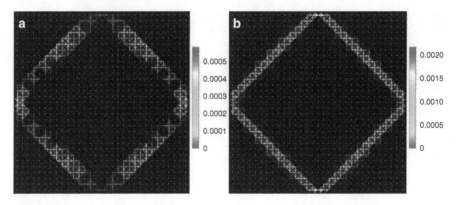

Fig. 5.8 Currents for a clean system with $t_f = 0.1$ and $V = 0.008$. (**a**) Currents in the c-electron subsystem and (**b**) Currents in the f-electron subsystem. The calculations were performed with spatially uniform mean-field parameters $\varepsilon_i = 0.025t$, $s = 0.3t$, and separate phonon couplings in the two bands equal to $\gamma_c = 10^{-7}t$ and $\gamma_f = 10^{-10}t$

to the retarded and advanced Green's functions for comparison with experimental quantities. Out of equilibrium, the lesser Green's function must be specified in addition to the retarded one, leading to a more complicated set of self-consistency relations [9].

The retarded and advanced Green's functions take the same form as in equilibrium (the Dyson equation):

$$\hat{G}_f^{R,A}(\omega) = \left\{ [\hat{g}_f^{R,A}(\omega)]^{-1} - \hat{s}\hat{g}_c^{R,A}(\omega)\hat{s} \right\}^{-1} \tag{5.3}$$

Here $\hat{g}$, $\hat{G}$, and $\hat{s}$ are matrices in the site indices of the square lattice. The condition on the f-electron occupation n_f is enforced through the calculation of the lesser Green's function, $\hat{G}_{ff}^<(\omega)$, given by

$$\hat{G}_{ff}^<(\omega) = \hat{g}_f^<(\omega) + \hat{g}_f^R(\omega)\hat{s}\hat{g}_c^R(\omega)\hat{s}\hat{G}_f^<(\omega) + \hat{g}_f^R(\omega)\hat{s}\hat{g}_c^<(\omega)\hat{s}\hat{G}_f^A(\omega)$$
$$+ \hat{g}_f^<(\omega)\hat{s}\hat{g}_c^A(\omega)\hat{s}\hat{G}_f^A(\omega) \tag{5.4}$$

After some algebra, one obtains

$$\hat{G}_{ff}^<(\omega) = \hat{G}_f^R(\omega)\hat{s}\hat{g}_c^<(\omega)\hat{s}\hat{G}_f^A(\omega) + \hat{G}_f^R(\omega)[\hat{g}_f^R(\omega)]^{-1}\hat{g}_f^<(\omega)[\hat{g}_f^A(\omega)]^{-1}\hat{G}_f^A(\omega) \tag{5.5}$$

Then self-consistency requires

$$n_f(\mathbf{r}) = 1 = \int_{-\infty}^{\infty} \frac{d\omega}{\pi} \mathrm{Im} G_{ff}^<(\mathbf{r},\mathbf{r},\omega) n_F(\omega) \tag{5.6}$$

5.5 Self-Consistency with Finite Bias

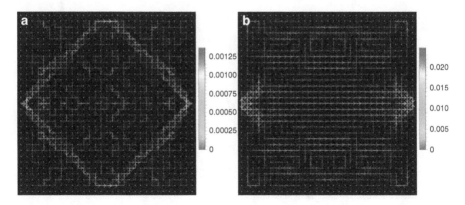

Fig. 5.9 Self-consistently calculated currents in the presence of finite bias, for (**a**) $V = 0.004t$ and (**b**) $0.3t$

We also have a self-consistency condition to determine the hybridization, s,

$$s(\mathbf{r}) = \frac{J}{2} \int_{-\infty}^{\infty} \frac{d\omega}{\pi} \mathrm{Im} G_{fc}^{<}(\mathbf{r}, \mathbf{r}, \omega) n_F(\omega) \tag{5.7}$$

where $n_F(\omega)$ is the Fermi distribution function. Here the dressed lesser cf Green's function is given by

$$\hat{G}_{fc}^{<}(\omega) = -\hat{g}_{cc}^{r}(\omega) \hat{s} \hat{G}_{ff}^{<}(\omega) - \hat{g}_{cc}^{<} \hat{s} \hat{G}_{ff}^{a}(\omega) \tag{5.8}$$

Results for the self-consistently calculated current patterns at $V = 0.004t$ and $0.3t$ are shown in Fig. 5.9. As expected, for small bias the results reproduce those of the non-self-consistent limit. However, with larger bias, the resulting current pattern is highly damped. We also plot the difference in the hybridizations between the non-equilibrium and equilibrium states, with $V = 0.1t, 0.2t$, respectively. This is shown in Fig. 5.10. We find clear indications of oscillatory behavior in the magnitude of the hybridization, emanating from the points where the leads are attached. In particular, some sites in the nanostructure actually have larger hybridization. Such behavior could potentially be confirmed in scanning tunneling spectroscopy experiments on heavy fermion systems, and if realized, would provide a dramatic example of the complex interplay of strongly correlated and non-equilibrium physics.

The correlation between increased voltage bias and the suppression of hybridization can be seen in a spatial plot of the difference in the hybridization at each site, obtained by subtracting from two high bias self-consistent solutions the hybridizations of one of the low bias cases. In Fig. 5.11a we show the difference in hybridization from the $V = 0.001t$ to the $V = 0.025t$ case, while Fig. 5.11b shows the same for $V = 0.075t$ as the upper value. The results are uniform in the y-direction due to the use of wide leads and periodic boundary conditions. While some sites experience an increase in hybridization, the majority have their hybridization

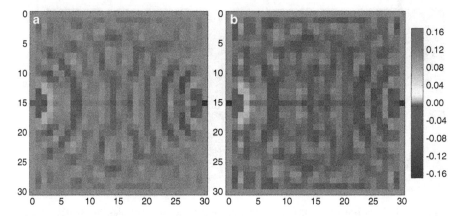

Fig. 5.10 Difference in the hybridization between the non-equilibrium and equilibrium states for (**a**) $V = 0.1t$ and (**b**) $V = 0.2t$

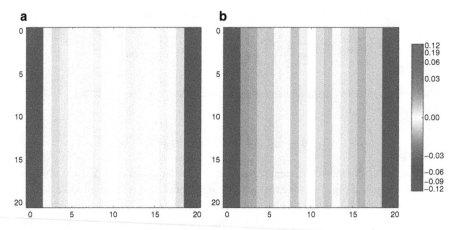

Fig. 5.11 Difference in hybridization, where values for the $V = 0.001t$ case were subtracted from the (**a**) $V = 0.025t$ and (**b**) $V = 0.075t$ cases

value suppressed, especially for those sites close to the leads. Thus, increasing the bias has the overall effect of decreasing the hybridization in the system.

To conclude, this chapter has examined the behavior of a nanoscale heavy fermion system out of equilibrium. In particular, we have shown how the presence of correlations between the conduction and localized f-electrons has a drastic impact on the pattern of charge currents flowing through the system. The changes on the current pattern induced by defects were also found to be influenced strongly by correlations. In particular, correlations appear to smooth out some of the modifications of the current in the vicinity of a defect, apart from the cross-like features that extend as far as 20 or more unit cells from the impurity. The addition of hopping in the f-band allows for an even greater range of behavior, and taking this step reveals the dominance of the f-electron states in the low energy dynamics. Finally, calculating the mean-field parameters self-consistently in the presence of a finite applied bias shows oscillations in the hybridization originating from where the leads are attached. Overall, as the bias (and therefore the current) increases, the average hybridization in the system is suppressed.

References

1. D.C. Langreth, Linear and nonlinear response theory with applications, in *Linear and Nonlinear Electron Transport in Solids*, ed. by J.T. Devreese, V.E.v. Doren, NATO Advanced Study Institutes Series, vol. 17 (Springer, New York, 1976), pp. 3–32. https://doi.org/10.1007/978-1-4757-0875-2_1
2. T. Can, H. Dai, D.K. Morr, Current eigenmodes and dephasing in nanoscopic quantum networks. Phys. Rev. B **85**(19), 195459 (2012)
3. L.V. Keldysh, Diagram technique for nonequilibrium processes. Zh. Eksp. Teor. Fiz. **47**, 1515 (1964). [Sov. Phys. JETP **20**, 1018 (1965)]
4. C. Caroli, R. Combescot, P. Nozieres, D. Saint-James, Direct calculation of the tunneling current. J. Phys. C Solid State Phys. **4**(8), 916 (1971)
5. J. Figgins, D.K. Morr, Defects in heavy-fermion materials: unveiling strong correlations in real space. Phys. Rev. Lett. **107**(6), 066401 (2011)
6. K.C. Nowack, E.M. Spanton, M. Baenninger, M. König, J.R. Kirtley, B. Kalisky, C. Ames, P. Leubner, C. Brüne, H. Buhmann, L.W. Molenkamp, D. Goldhaber-Gordon, K.A. Moler, Imaging currents in HgTe quantum wells in the quantum spin Hall regime. Nat. Mater. **12**(9), 787–791 (2013)
7. T. Can, D.K. Morr, Atomic resolution imaging of currents in nanoscopic quantum networks via scanning tunneling microscopy. Phys. Rev. Lett. **110**(8), 086802 (2013)
8. Z. Bihary, M.A. Ratner, Dephasing effects in molecular junction conduction: an analytical treatment. Phys. Rev. B **72**(11), 115439 (2005)
9. H. Haug, A.-P. Jauho, *Quantum Kinetics in Transport and Optics of Semiconductors*. Solid-State Sciences, vol. 123 (Springer, Berlin, 2008)

Chapter 6
Charge and Spin Currents in Nanoscale Topological Insulators

6.1 Introduction

Topological insulators (TIs) have generated sustained interest for nearly a decade. These materials are characterized by the presence of topological invariants: global properties of the system that are quantized and therefore cannot be changed under smooth deformations of the underlying Hamiltonian without closing the gap at the Fermi level. An example is the Berry phase obtained by integrating the gradient of the Bloch wavefunction of a crystalline insulator around a closed loop in the Brillouin zone [1, 2]. The vacuum is trivially an insulator, with a Berry phase of zero. If the system possesses a non-zero Berry phase (say equal to one), then its edge must be conducting. This is because the quantized Berry phase is forced to take integer values, and thus it cannot go from a value of one inside the material to zero in the vacuum while remaining an insulator. Over the years, much theoretical work has been done on the classification of the different topological states and their possible realization in experiment [3, 4]. Experiments, on the other hand, have naturally focused on confirming the various theoretical predictions. Throughout this process, a partial justification for the work in this field has been found in proposed applications, such as spintronics and quantum computing. As work shifts away from basic questions and toward applications, new questions arise about the various models of topological insulators and their realizations in materials. As an example, consider the use of TIs in spintronics, where the generation of spin-polarized currents is a central concern. To spur development in this direction, it is necessary to suggest concrete procedures by which spin-polarized currents can be realized.

6.2 Model

Although several different models of topological insulators exist, for the sake of definiteness consider the one due to Kane and Mele [5] on the two-dimensional honeycomb lattice (the crystal structure of graphene).

$$H = -t \sum_{<\mathbf{r},\mathbf{r'}>,\alpha} c^\dagger_{\mathbf{r},\alpha} c_{\mathbf{r'},\alpha} + i\Lambda_{SO} \sum_{<<\mathbf{r},\mathbf{r'}>>,\alpha,\beta} v_{\mathbf{r},\mathbf{r'}} c^\dagger_{\mathbf{r},\alpha} \sigma^z_{\alpha\beta} c_{\mathbf{r'},\beta}$$

$$- t_l \sum_{\mathbf{r},\mathbf{r'},\alpha} (d^\dagger_{\mathbf{r},\alpha} c_{\mathbf{r'},\alpha} + h.c.) + H_l \qquad (6.1)$$

The first term gives the ordinary hopping amplitude for electrons between nearest-neighbor sites, while the second describes the nearest-nearest-neighbor hopping due to spin-orbit coupling. Here, $v_{\mathbf{r},\mathbf{r'}} = -v_{\mathbf{r'},\mathbf{r}} = \pm 1$ and $\sigma^z_{\alpha\beta}$ is a Pauli matrix. The sign of $v_{\mathbf{r},\mathbf{r'}}$ is determined by the direction of the hopping around the honeycomb: positive for counterclockwise and negative for clockwise motion. This term is essential for producing the non-trivial topological behavior. The third term gives the coupling between the leads and the system, whereas the fourth term describes the Hamiltonian of the leads. In the following, the leads are modeled via a continuous and flat density of states, as appropriate for a macroscopic metallic system. Note that Eq. (6.1) differs from that considered by Kane and Mele, in that it neglects the Rashba coupling for simplicity.

In the following, the spatially-resolved currents through the nanoscale TI are calculated using the Keldysh Green's function method [6, 7], which was briefly introduced in Chap. 5 and is discussed in detail in Appendix A. Apart from the different Hamiltonian in the case of the Kondo lattice as opposed to the TI, the method for computing the currents is the same in both systems.

6.3 Polarized Spin Currents

We now turn to the demonstration of highly spin-polarized currents in nanoscale TIs using magnetic defects. This is in fact the first theoretical proposal showing the creation of highly spin-polarized currents in these systems. The possibility of creating such currents is found to be robust against variations in the model parameters, such as size and shape of the TI, the width of the leads, and the strengths of the spin-orbit couplings and impurity magnetic scatterers. This will be demonstrated in Sect. 6.8. For now, consider a nanoscale TI whose dimensions along the armchair and zigzag edges are $N_a = 9$ and $N_z = 15$, respectively. The TI is connected to two narrow, metallic leads at L and R, as shown schematically in Fig. 6.1.

The finite size of the system under consideration implies the discreteness of its energy levels, which appear as sharp peaks in the density of states, broadened by an electronic dephasing due to the coupling to the leads. The numerically calculated local density of states at the site of the TI connected to the left lead, $N_\sigma(\mathbf{r} = L, E)$,

6.3 Polarized Spin Currents

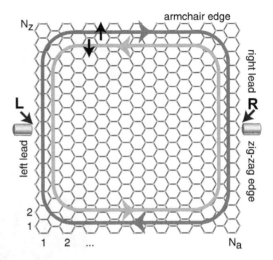

Fig. 6.1 Schematic drawing of the spin-resolved current patterns in a two-dimensional topological insulator on the honeycomb lattice [8]

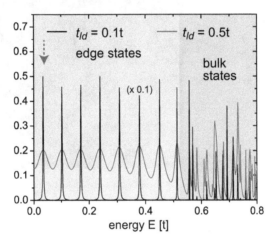

Fig. 6.2 Energy dependence of the local density of states(LDOS), $N_{\uparrow,\downarrow}$, at the site L attached to the left lead, for two different values of the coupling t_l. The purple background indicates edge states, while green indicates bulk states. Without magnetic defects, the system is particle-hole symmetric, and the LDOS is therefore presented for $E > 0$ only. The states are broadened by an electron dephasing of $\delta = 10^{-5}t$ and the electronic hoppings are $t = 1.0$ and $t_l = \Lambda_{SO} = 0.1t$. For visualization purposes, the LDOS for $t_l = 0.1t$ has been multiplied by 0.1

is shown in Fig. 6.2. The states with a purple background below the spin-orbit gap are edge states, as evidenced by their associated current patterns (see below). Those with the green background above the spin-orbit gap are higher energy edge states that exist outside of the spin-orbit gap of magnitude $\Delta_{SO} = 3\sqrt{3}\Lambda_{SO}$. Similar results were also found in a cylindrical geometry [9].

One may choose a particular state to carry the current by gating the system capacitively (Fig. 6.3). For a state at energy E_i, applying a gate voltage $V_g = E_i/e$

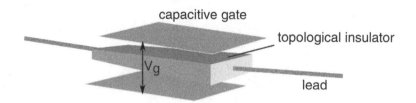

Fig. 6.3 Schematic drawing of a TI illustrating the capacitive gating of the system to select states for transport

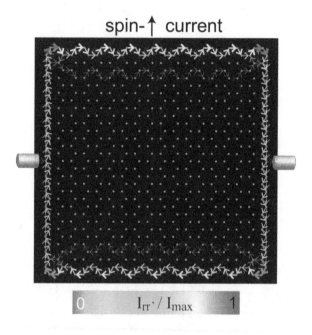

Fig. 6.4 Spatial pattern of the spin-↑ current, $I_{rr'}^{\uparrow}$, carried by the lowest energy edge state at $E_1 = 0.0342t$ (see blue dashed arrow in Fig. 6.2) for coupling $t_l = 0.1t$. This state is accessed by applying a gate voltage $V_g = E_1/e$ to the TI. Note the existence of a quantum mechanical backflow branch along the TI's lower edge where a current flows opposite to the applied voltage bias

brings it to the Fermi level, allowing the current to flow. For a system free of defects and impurities, the resultant spin-↑ current pattern for the state at $E_1 = 0.0342t$ (indicated by the blue dashed arrow in Fig. 6.2) is shown in Fig. 6.4. As expected for an edge state, the current is strongly confined to the perimeter of the sample. In addition to the ordinary flow along the top edge from source to sink, there is quantum-mechanical backflow along the bottom edge [10]. This leads to a circulating current pattern with a much greater magnitude than the outgoing current. As the gate voltage V_g is increased, it is found that the edge states penetrate further into the bulk along the zigzag edge [11], as shown in Fig. 6.5. Since the system-lead coupling destroys the electronic phase coherence and thus breaks the macroscopic time-reversal symmetry [2], the current pattern is dependent on the coupling strength. In particular, for a large coupling of $t_l = 0.5t$, the backflow is suppressed and the current for spin-↑ and spin-↓ electrons is confined to the upper and lower branches, respectively, as shown in Fig. 6.6.

6.3 Polarized Spin Currents

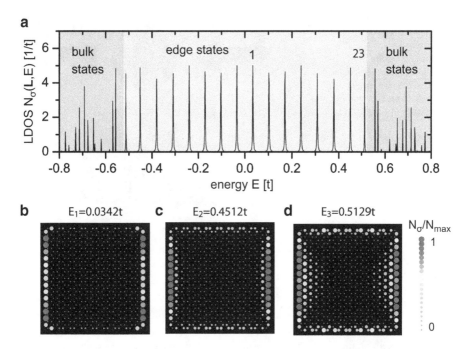

Fig. 6.5 Decay of edge states into the bulk, where (**a**) shows the local density of states at the site attached to the left lead and (**b**)–(**d**) show the spatial pattern of the edge states at the energies E_1–E_3 indicated in (**a**). The high energy edge states decay further into the bulk along the zigzag edge

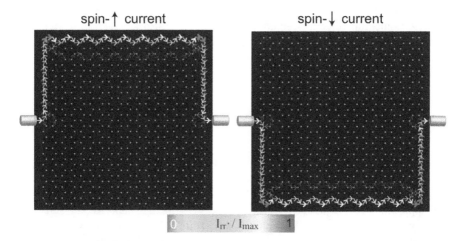

Fig. 6.6 Spatial pattern of the spin-↑ current, $I_{rr'}^{\uparrow}$, and the spin-↓ current, $I_{rr'}^{\downarrow}$, respectively, carried by the edge state at $E_1 = 0.0335t$ for coupling $t_l = 0.5t$. Color (see legend) and thickness of the arrows represent the magnitude of the normalized current $I_{rr'}^{\sigma}/I_{max}^{\sigma}$ (the same normalization is used for both subplots)

Fig. 6.7 Schematic drawing of potential scattering in a TI. Electrons are not backscattered due to spin-momentum locking and the absence of a mechanism for transitions between the spin bands

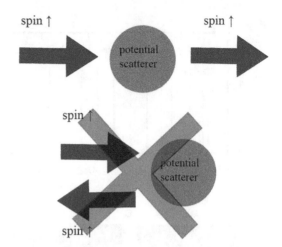

6.4 Non-magnetic Defects

When non-magnetic defects (such as localized potential scatterers) are added to the system, the energies of the edge states are subject to modification. This also leads to local changes in the resulting current pattern, since a strong repulsive potential at a site leads to a suppression of the current through it. However, potential scatterers are not spin-dependent, thus maintaining the time-reversal symmetry of the TI and preventing backscattering from one spin channel to the other. That is, the potential scatterer does not introduce any terms into the Hamiltonian that allow for transitions between the spin-↑ and spin-↓ bands (Fig. 6.7). Thus, the spin projection and the momentum direction remain locked in this scenario, so that backscattering is impossible. These results are demonstrated in Fig. 6.8 and reveal the marked contrast between impurities in topological materials and, for instance, the correlated but non-topological Kondo lattice discussed in Chap. 5.

6.5 Magnetic Defects

In order to obtain a net spin-polarization of the current through the TI, it is necessary to introduce magnetic defects that break the time-reversal symmetry inside the system. Here we introduce the impurities as static, spin-dependent scatterers. Such an approach will be justified if the magnetic moment is not Kondo screened. However, it is known that the Kondo temperature [12, 13] can be suppressed in various ways, for instance, by a lack of edge states near the Fermi energy [14], by using large-spin defects, or by local magnetic fields [15]. On the other hand, the topological properties of the system can survive all the way up to room temperatures [16]. Hence, there will be experimentally accessible regimes in which the magnetic impurities can be treated as static [17].

6.5 Magnetic Defects

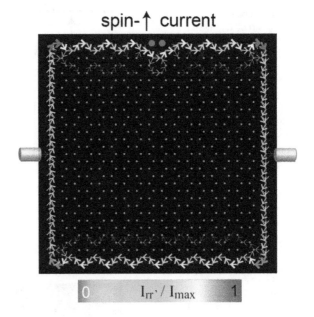

Fig. 6.8 Spatial pattern of the spin-↑ current, $I^{\uparrow}_{\mathbf{rr}'}$, carried by the lowest energy edge state at $E_1 = 0.031t$ in the presence of two potential defects (locations indicated by red dots) with scattering strength $U_0 = 10t$ and $t_l = 0.1t$

In light of this, the Hamiltonian for the point-like magnetic impurities can be written as

$$H_M = \sum_{\mathbf{R}} J_z S^z_{\mathbf{R}}(c^\dagger_{\mathbf{R},\uparrow}c_{\mathbf{R},\uparrow} - c^\dagger_{\mathbf{R},\downarrow}c_{\mathbf{R},\downarrow}) + J_\pm(S^+_{\mathbf{R}}c^\dagger_{\mathbf{R},\downarrow}c_{\mathbf{R},\uparrow} + S^-_{\mathbf{R}}c^\dagger_{\mathbf{R},\uparrow}c_{\mathbf{R},\downarrow})$$

(6.2)

This Hamiltonian includes two distinct types of magnetic scattering. The first term with coupling constant J_z represents an Ising-type defect and is akin to the Zeeman effect of an external magnetic field in that it splits the spin degeneracy by raising (lowering) the energy of the spin-↑ (spin-↓) state (for $J_z > 0$). The second term with coupling constant $J_\pm$ produces a spin-flip scatterer which allows the electrons to hop between the two spin bands. We proceed to discuss the two cases in turn.

6.5.1 Ising-Type Magnetic Defects

The separation of the spin-↑ and spin-↓ bands by an Ising-type defect ($J_z \neq 0$, $J_\pm = 0$) placed at the edge of the sample is clearly revealed in a numerical calculation of the spin-resolved local density of states at site L (Fig. 6.9). By gating the system, as discussed above, one may select a particular state for transport, which now only contains electrons of one spin projection. Consider, for instance, the state

Fig. 6.9 Local density of states, $N_\sigma(L, E)$, in a TI without (black line) and with (red line for spin-↑, green for spin-↓) a magnetic defect (red dot in Fig. 6.10) with Ising-type symmetry: $J_z S = 5t$ and $t_l = 0.1t$. The blue arrow indicates state used for transport in Fig. 6.10

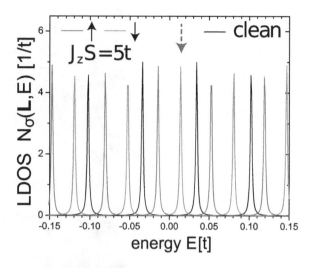

Fig. 6.10 Spatial pattern of the charge current, $I_{rr'}^c$, carried by the lowest energy edge state at $E_1 = 0.0142t$ [see blue dashed arrow in Fig. 6.9] for a system with an Ising-type defect

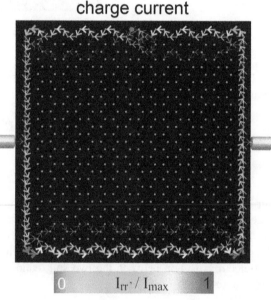

$E_1 = 0.0142t$, indicated by the blue arrow in Fig. 6.9. For the weak lead-system coupling limit, where the width of the states is much smaller than their separation in energy, calculating the current through the system that is carried by this state yields the current pattern shown in Fig. 6.10. Similar to the case of potential scatterers, the current pattern is locally modified near the defect (indicated by the red dot in Fig. 6.10). However, a calculation of the spin-polarization $\eta_\downarrow$ reveals that the outgoing current is 98% spin-↓polarized. Thus, Ising-type defects are highly efficient at creating spin-polarized currents in the presence of weak coupling to the leads.

6.5.2 Spin-Flip-Type Magnetic Defects

For the case of strong lead-system coupling, spin-flip scatterers ($J_\pm \neq 0, J_z = 0$) provide a mechanism for achieving large spin-polarizations. When an electron of a given spin projection scatters against a spin-flip defect, it is transferred to the opposite spin band. This has striking consequences for the case of a topological insulator, in which the spin is correlated with the direction of motion of the electron around the edge. An electron that encounters a spin-flip defect is scattered, reversing its direction—the defect effectively blocks current flow through the branch in which it is located. This is shown in Fig. 6.11, where the particular state chosen is indicated by the blue arrow in Fig. 6.12. With the spin-flip defect located in the top branch, the spin-↑ current that would naturally flow along that path is scattered into spin-↓ current, which travels the opposite direction around the edge. This combines with the spin-↓ current that enters from the left lead, and exits the system through the right lead. The behavior can be partially understood from the local density of states, shown for this case in Fig. 6.12. Here the states for spin-↑ and spin-↓ are strongly overlapping, so that electrons can be scattered between the spin bands. On the other hand, if the density of states of the spin-↑ (spin-↓) band vanished at the specified gate voltage, there would be no states for the electrons to scatter out of (into). Since the spin-↑ current is blocked from reaching the right lead, the net current is spin-↓ polarized with $\eta_\downarrow = 96.5\%$. Because of the overlapping of the edge states, this result is relatively insensitive to variations in the gate voltage V_g. Examining the total charge current $I_{out}^c = \sum_\alpha I_{out}^\alpha$, one notices that the counter-propagating spin-↑ and spin-↓ currents in the upper branch cancel, leaving a charge current only along the bottom branch, as shown in Fig. 6.13.

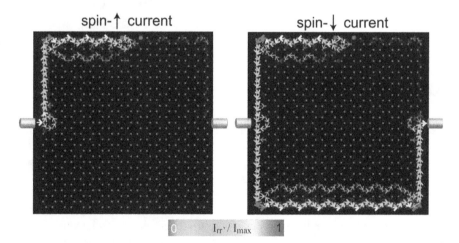

Fig. 6.11 $I_{rr'}^\uparrow$, and $I_{rr'}^\downarrow$ carried by the edge state at $E_1 = 0.0175t$ [see blue dashed arrow in Fig. 6.12] for $t_l = 0.5t$

Fig. 6.12 Local density of states, $N_\sigma(L, E)$, for a TI containing two magnetic defects [red dots in Fig. 6.11] with xy symmetry, $J_\pm S = 5t$ and $t_l = 0.5t$

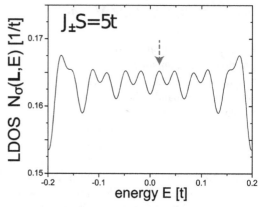

Fig. 6.13 Spatial pattern of $I^c_{rr'}$ carried by the edge state at $E_1 = 0.0175t$ (see blue dashed arrow in Fig. 6.12) for $t_l = 0.5t$

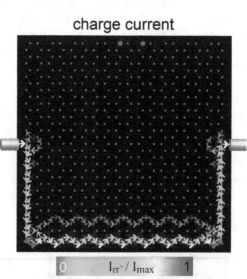

6.6 Heisenberg Defects and Spin Diodes

Sections 6.5.1 and 6.5.2 considered the cases in which the magnetic impurity had either Ising or xy symmetry. However, if $J_z = J_\pm$, the resulting defect is isotropic in spin space and can be represented by $J\mathbf{S} \cdot \boldsymbol{\sigma}$, where $\boldsymbol{\sigma}$ is a vector of the Pauli matrices. Such Heisenberg symmetry defects can be effective for generating spin-polarizations over a broad range of lead-system couplings, since one or the other of the mechanisms discussed above is active in a given regime. Furthermore, this situation allows for the creation of "spin diodes" in the following manner. Consider a system with $J_z = J_\pm = 5.0t$ and $t_l = 2.5t$ and two Heisenberg symmetry defects in the upper branch (as shown in Fig. 6.14). The corresponding spin-polarizations $\eta_{\uparrow,\downarrow}$ are shown in Fig. 6.15. When the gate voltage is set to $V_{g,1}$, one finds that for forward bias ΔV the spin-$\downarrow$ polarization $\eta_\downarrow$ is large, whereas

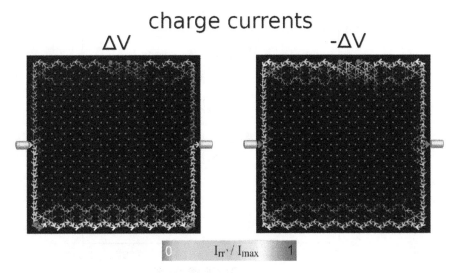

Fig. 6.14 Spatial pattern of $I^c_{rr'}$ with a Heisenberg-type defect for forward bias ΔV and backward bias $-\Delta V$ at $V_{g,1}$

$\eta_\uparrow$ is small. This occurs because spin-flip scattering off the defects in the upper branch suppresses the outgoing spin-$\uparrow$ current. However, when the bias is reversed the two polarizations become nearly equal (Fig. 6.15). In this case, since the current now flows in the opposite direction, the spin-$\downarrow$ current is blocked by the defect and $\eta_\downarrow$ is correspondingly reduced. However, this reduction is partially offset by the fact that the density of states for spin-$\downarrow$ electrons is greater at the energy $V_{g,1}$, in consequence of the Ising component of the defect, which is responsible for the splitting of the spin-polarized states. Note that the current is larger because there are more states available for transport at this energy. While the spin-polarization is therefore changed due to a bias reversal, the magnitude of the charge current is unaffected. Thus, the system behaves as a spin diode, with a polarization that can be turned on and off by switching the bias direction.

Although the magnitude of the charge current stays the same, the spatial pattern is notably different between the forward and backward bias cases. In the former, the charge current travels predominantly along the bottom edge of the TI, reminiscent of the xy symmetry defect. With the backward bias ($-\Delta V$), the current travels equally in both the upper and lower branches (which is also reflected in the fact that $\eta_\uparrow = \eta_\downarrow$). Thus there is a correlation between the presence of a net spin-polarization (which exists for forward, but not for backward, bias) and the spatial pattern of the charge current. This implies that a net spin-polarization can be detected by imaging the charge currents in the system [18] (see [7] of Chap. 5).

At other energies than $V_{g,1}$ the behavior of the system under bias reversal is different. For instance, at $V_{g,2}$ the density of states is equal for spin-$\uparrow$ and spin-$\downarrow$ electrons, and for forward bias $\eta_\downarrow > \eta_\uparrow$. Upon bias reversal we find $\eta_\downarrow < \eta_\uparrow$,

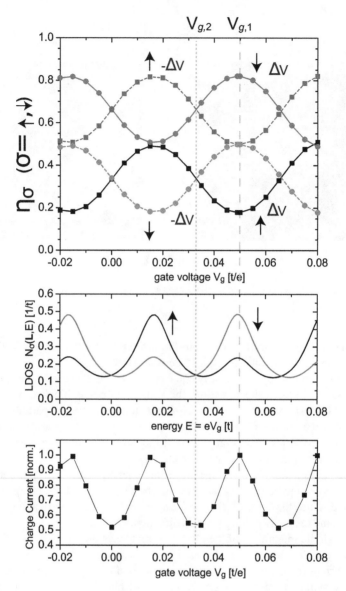

Fig. 6.15 TI containing two magnetic defects of Heisenberg symmetry with $J_z S = J_\pm S = 5t$ and $t_l = 0.275t$. (Top panel) $\eta_{\uparrow,\downarrow}$ as a function of V_g for forward, ΔV ($\eta_\uparrow$: black line, $\eta_\downarrow$: red line), and backward bias, $-\Delta V$ ($\eta_\uparrow$): blue dashed line, $\eta_\downarrow$: green dashed line). (Center panel) N_σ ($E = eV_g$) at L. (Bottom panel) Total normalized charge current $I^c_{out}(V_g)/I^c_{max}$ with $I^c_{max} = \max_{V_g}(I^c_{out})$

so that spin current has opposite polarization. The magnitude of the difference $|\eta_\uparrow - \eta_\downarrow|$, however, remains unchanged, although the direction of the current flow is now reversed. As before, the magnitude of the charge current remains the same, but unlike the previous case, the spatial pattern of the charge current is not sensitive to the bias reversal—larger current always flows along the bottom branch. Since the charge current pattern no longer reflects the spin-polarization, a spin-polarized experimental probe would be needed to detect this effect.

6.7 Interface with Ferro- and Antiferromagnets

Another possibility for realizing highly spin-polarized currents is to interface the TI on the nano- or mesoscale with a ferro- or antiferromagnet. Suppose a magnet is placed in contact with the TI along the top edge. Assuming the magnet is in a topologically trivial insulating state (Chern number equal to zero), a conducting surface state will still exist along the upper edge of the TI. However, the electrons in the TI will experience the influence of the adjoining magnet through a proximity effect. If we interface the TI along the top edge with a ferromagnet with an easy-plane parallel to the surface, this can be modeled as a row of xy-symmetry magnetic impurities that scatter electrons at the relevant sites.

In experiment one would expect some amount of disorder to exist along the interface, which would scatter the electrons traveling along the edge. This can be modeled by introducing random vacancies in the ferromagnet, i.e. sites where the coupling $J_\pm = 0$. Performing the calculations for this scenario, we find that the spin-↑ current travels along the upper edge until it encounters the first vacancy, whereupon it is strongly scattered, as seen in Fig. 6.16. The spatial pattern strongly resembles the case of a single xy symmetry defect, Fig. 6.11, with the magnetic impurity in that case being replaced by the hole in the ferromagnet in the present situation. The resulting spin-polarization is still very high, with $\eta_\downarrow = 0.99$. One can model the Interface of the TI with an antiferromagnet by changing the sign of the scattering potential between neighboring sites. This also yields a strong spin-polarization of $\eta_\downarrow = 0.90$, along with the current patterns of Fig. 6.17. One notices that the decay length of the spin-↑ current along the top edge is much

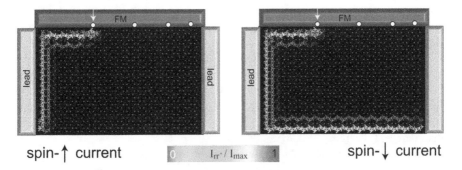

Fig. 6.16 Spatial pattern of (**a**) $I_{rr'}^\uparrow$ and (**b**) $I_{rr'}^\downarrow$ for a TI interfaced with a disordered ferromagnet ($J_\pm S = 5t$; white circles indicate vacancies with $J_\pm S = 0$). For this system, $N_a = 14$, $N_z = 15$, $t_l = 0.5t$

Fig. 6.17 Spatial pattern of (**a**) $I_{rr'}^{\uparrow}$ and (**b**) $I_{rr'}^{\downarrow}$ for a TI interfaced with a antiferromagnet ($J_{\pm}S = \pm 5t$; sign of $J_{\pm}S$ varies between neighboring sites). For this system, $N_a = 14$, $N_z = 15$, $t_l = 0.5t$

greater than for the ferromagnet or the single magnetic defect cases. The slow decay of the current indicates that each pair of neighboring anti-aligned spins produces a small amount of the total spin-polarization. This is confirmed by examining systems of increasing length along the armchair edge. While the $N_a = 14$ system has $\eta_{\downarrow} = 0.90$, increasing the length results in $\eta_{\downarrow} = 0.96$ for $N_a = 18$ and $\eta_{\downarrow} = 0.99$ for $N_a = 25$. The fact that the spin-polarization of these hybrid nanostructures remains very high with increasing size suggests the effects will persist in the meso- and macroscales as well.

6.8 Robustness of the Spin-Polarized Currents

For proposed applications, it is crucial that the phenomenon of spin-polarization explored in this chapter be robust in the variety of conditions that are likely to be realized in experiments. If the results found above depended heavily on the fine tuning of model parameters, it would be difficult to obtain them in realistic systems where it is challenging, if not impossible, to control various system parameters. Hence, it is important to examine the robustness of the proposed spin-polarization effects under the variation of model parameters, as done in the following section for edge disorder, the system size and geometry, the spin-orbit coupling, the width of the leads, and the strength of magnetic scattering.

Beginning with possibility of disorder along the edge of the system, consider a TI in which 30% of edge sites are randomly removed (Fig. 6.18) containing two magnetic defects of xy symmetry. While in such a TI, the spatial patterns of the spin-$\uparrow$ and spin-$\downarrow$ currents are more disordered (Fig. 6.19); the maximum spin-polarization (as a function of V_g) of $\eta_{\downarrow} = 0.975$ is similar to that of the non-disordered TI where $\eta_{\downarrow} = 0.965$.

To show that the spin-polarization does not rely on a specific system size or geometry, consider a system with $N_a = 14$, $N_z = 13$, which leads to an aspect ratio N_a/N_z that is considerably different than the case considered previously. The importance of considering different aspect ratios lies in the fact that, for

6.8 Robustness of the Spin-Polarized Currents

Fig. 6.18 Schematic of nanoscale TI with disordered edges

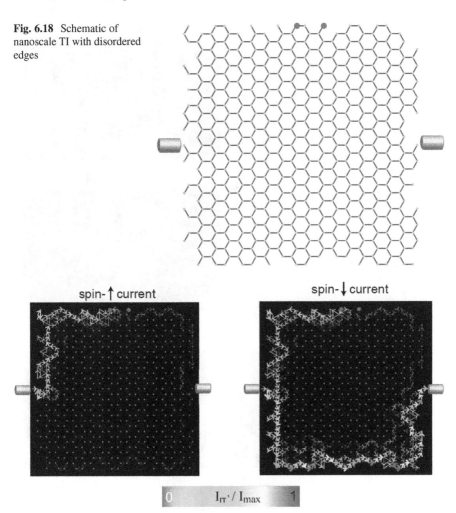

Fig. 6.19 Spin up and down currents through a TI with disordered edges and an xy defect

non-topologically-protected nanoscale networks, the aspect ratio has been found to have a profound influence on the resultant current patterns [10]. As to be expected, the electronic structure of this finite-size system is changed relative to the previous case—the states have moved in energy, consistent with the different number of sites and changed geometry. Placing an Ising-symmetry defect of magnitude $J_z S = 5t$ in the upper branch, as done above, and gating the system to select the spin-$\downarrow$ polarized state at $E = 0.0125t$ for transport, one finds a current pattern very similar to the one discussed above in Sect. 6.5. The density of states and current pattern are shown in Fig. 6.20a,b. More importantly, the spin-polarization of the outgoing current is 98.9%, almost identical to what was found above (Sect. 6.5.1). The same conclusion is also found to hold for defects with xy symmetry, as displayed in Fig. 6.20c–f. In

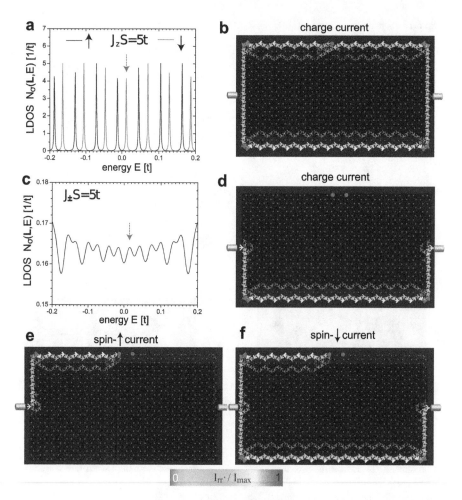

Fig. 6.20 For a TI with $N_a = 14$, $N_z = 13$, (**a**) local density of states and (**b**) charge current pattern in the presence of a defect with Ising symmetry ($J_z S = 5t$). (**c**) Local density of states, (**d**) charge current, (**e**) spin-↑ current, and (**f**) spin-↓ current in the presence of a defect with xy symmetry ($J_\pm S = 5t$)

this case, the edge states are broad enough that the gate voltage need not be changed relative to the earlier system geometry. Selecting again the state at $E = 0.0175t$, one finds similar current patterns as in Sect. 6.5, as well as a very similar overall spin-polarization of 96.2%.

The fascinating properties of topological insulators depend crucially on the presence of spin-orbit coupling in those systems [1]. Any proposal to generate spin-polarization from a topological insulator will require some amount of spin-orbit coupling for its realization, but ideally it should not rely on coupling strengths which are excessively large, as these will be difficult to obtain in practice. It is therefore important to check that the polarization can be produced with weaker values of

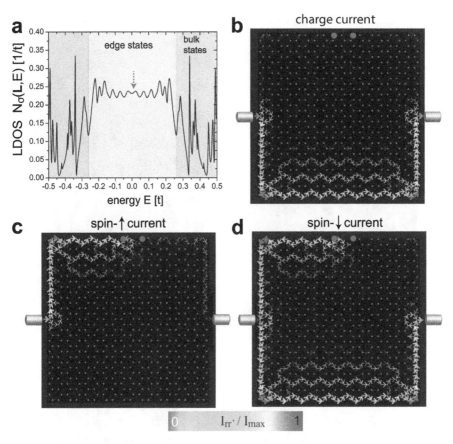

Fig. 6.21 (a) Local density of states, (b) charge currents, (c) spin-↑ currents, and (d) spin-↓ currents for a TI with $N_a = 14$, $N_z = 13$ and a magnetic defect [red dot] with xy symmetry, $J_{\pm}S = 5t$, $t_l = 0.1t$, and $\Lambda_{SO} = 0.05t$

the spin-orbit coupling than were employed above. Indeed, if one uses a coupling strength of $\Lambda_{SO} = 0.05t$ (half the original value), one finds only a slight reduction in the outgoing spin-polarization from 96.3% to 94.6%. As expected, the resulting spin-orbit gap shrinks as well. However, the spatial patterns of the current remain similar to those found with the large coupling strength, though the currents penetrate further into the bulk of the system, as seen in Fig. 6.21.

The result of high spin-polarization ought to be insensitive not only to the details of the TI, but to those of the leads as well. It may be possible to fabricate atomically-sharp leads using STM tips, but other experimental setups will have difficulty achieving the same level of precision. For instance, typical quantum point contacts possess diameters on the order of 10 nm [19]. To investigate the possibility that the lead geometry could influence the net spin-polarization, we attach wide leads to the TI and calculate the current. This is shown in Fig. 6.22. In Fig. 6.22a,b,c the spin-↑,

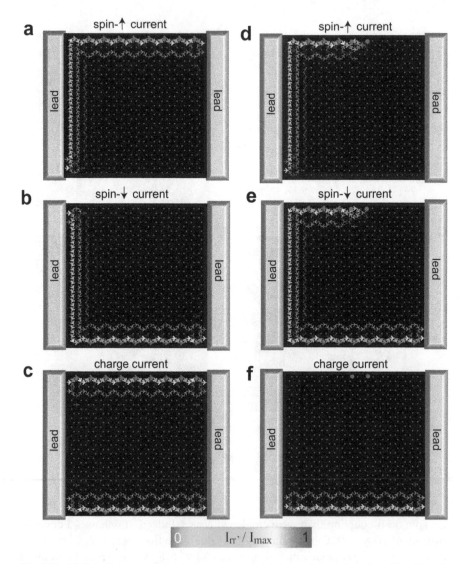

Fig. 6.22 (**a**) Spin-↑ current, (**b**) spin-↓ current, and (**c**) charge current in a clean TI with wide leads. (**d**) Spin-↑ current, (**e**) spin-↓ current, and (**f**) charge current in a TI with wide leads and two defects of xy-symmetry

spin-↓, and charge currents are shown for the case of a clean system. A somewhat surprising result is that the spin currents predominantly enter the system through a single site (the bottom left in the case of spin-↑). This leads to a cancellation in the charge current along the left edge, ensuring that the currents travel the shortest path possible from the source to the sink.

As a final check of the robustness of the spin-polarization, we consider variations in the magnetic scattering strength. The spin-polarization obtained using two defects

6.8 Robustness of the Spin-Polarized Currents

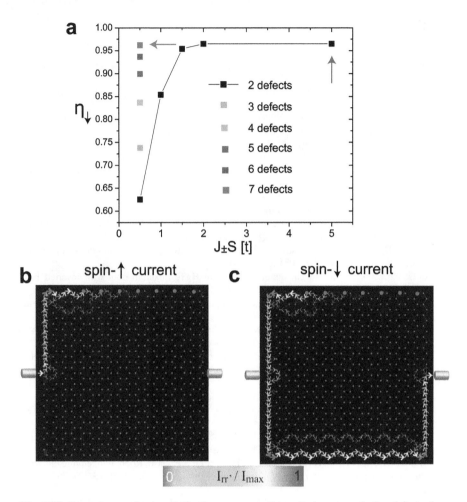

Fig. 6.23 Dependence of spin-polarization on magnetic scattering strength for defect of xy symmetry. (**a**) spin-polarization $\eta_\downarrow$ as a function of $J_\pm S$ for different numbers of defects along the top edge of the system. (**b**) Spin-↑ and (**c**) Spin-↓ current patterns for a system with 7 defects

of xy symmetry is shown in Fig. 6.23 as a function of scattering strength. The blue arrow in Fig. 6.23a shows the case investigated earlier in Figs. 6.11, 6.12, and 6.13. The polarization $\eta_\downarrow$ only begins to decrease appreciably for $J_\pm S <\approx 1.5t$. For $J_\pm S = 0.5t$, the system with two defects has a polarization of only $\eta_\downarrow = 0.625$, close to the unpolarized limit of $\eta_\downarrow = 0.5$. However, one can restore the high spin-polarization in this case simply by adding additional defects, as shown in Fig. 6.23a. For a system with 7 defects (red arrow in Fig. 6.23a), one obtains $\eta_\downarrow = 0.962$, close to the two defect case with large $J_\pm S$. The spin-resolved current patterns in this case are shown in Fig. 6.23b,c, which is qualitatively similar to the results of Fig. 6.11, but with a larger decay length of the spin-↑ current along the top edge.

References

1. M.Z. Hasan, C.L. Kane, Colloquium: topological insulators. Rev. Mod. Phys. **82**(4), 3045–3067 (2010)
2. X.-L. Qi, S.-C. Zhang, Topological insulators and superconductors. Rev. Mod. Phys. **83**(4), 1057–1110 (2011)
3. A. Altland, M.R. Zirnbauer, Nonstandard symmetry classes in mesoscopic normal-superconducting hybrid structures. Phys. Rev. B **55**(2), 1142–1161 (1997)
4. S. Ryu, A.P. Schnyder, A. Furusaki, A.W.W. Ludwig, Topological insulators and superconductors: tenfold way and dimensional hierarchy. New J. Phys. **12**(6), 065010 (2010)
5. C.L. Kane, E.J. Mele, Z_2 topological order and the quantum spin hall effect. Phys. Rev. Lett. **95**(14), 146802 (2005)
6. L.V. Keldysh, Diagram technique for nonequilibrium processes. Zh. Eksp. Teor. Fiz. **47**, 1515 (1964). [Sov. Phys. J. Exp. Theor. Phys. **20**, 1018 (1965)]
7. C. Caroli, R. Combescot, P. Nozieres, D. Saint-James, Direct calculation of the tunneling current. J. Phys. C Solid State Phys. **4**(8), 916 (1971)
8. J.S. Van Dyke, D.K. Morr, Controlling the flow of spin and charge in nanoscopic topological insulators. Phys. Rev. B **93**(8), 081401 (2016)
9. K. Chang, W.-K. Lou, Helical quantum states in HgTe quantum dots with inverted band structures. Phys. Rev. Lett. **106**(20), 206802 (2011)
10. T. Can, H. Dai, D.K. Morr, Current eigenmodes and dephasing in nanoscopic quantum networks. Phys. Rev. B **85**(19), 195459 (2012)
11. E. Prada, G. Metalidis, Transport through quantum spin Hall insulator/metal junctions in graphene ribbons. J. Comput. Electron. **12**(2), 63–75 (2012)
12. C. Wu, B.A. Bernevig, S.-C. Zhang, Helical liquid and the edge of quantum spin hall systems. Phys. Rev. Lett. **96**(10), 106401 (2006)
13. J. Maciejko, C. Liu, Y. Oreg, X.-L. Qi, C. Wu, S.-C. Zhang, Kondo effect in the helical edge liquid of the quantum spin hall state. Phys. Rev. Lett. **102**(25), 256803 (2009)
14. E. Rossi, D.K. Morr, Spatially dependent Kondo effect in quantum corrals. Phys. Rev. Lett. **97**(23), 236602 (2006)
15. D. Rugar, R. Budakian, H.J. Mamin, B.W. Chui, Single spin detection by magnetic resonance force microscopy. Nature **430**(6997), 329–332 (2004)
16. A. Dankert, J. Geurs, M.V. Kamalakar, S. Charpentier, S.P. Dash, Room temperature electrical detection of spin polarized currents in topological insulators. Nano Lett. **15**(12), 7976–7981 (2015)
17. Y.L. Chen, J.-H. Chu, J.G. Analytis, Z.K. Liu, K. Igarashi, H.-H. Kuo, X.L. Qi, S.K. Mo, R.G. Moore, D.H. Lu, M. Hashimoto, T. Sasagawa, S.C. Zhang, I.R. Fisher, Z. Hussain, Z.X. Shen, Massive Dirac fermion on the surface of a magnetically doped topological insulator. Science **329**(5992), 659–662 (2010)
18. K.C. Nowack, E.M. Spanton, M. Baenninger, M. König, J.R. Kirtley, B. Kalisky, C. Ames, P. Leubner, C. Brüne, H. Buhmann, L.W. Molenkamp, D. Goldhaber-Gordon, K.A. Moler, Imaging currents in HgTe quantum wells in the quantum spin Hall regime. Nat. Mater. **12**(9), 787–791 (2013)
19. G. Deutscher, Point contact spectroscopy in strongly correlated systems, in *Strongly correlated systems*, ed. by A. Avella, F. Mancini. Springer Series in Solid-State Sciences, vol. 180 (Springer, Berlin, 2015), pp. 111–135. https://doi.org/10.1007/978-3-662-44133-6_4

Chapter 7
Conclusion

This work has explored aspects of the complex behavior that arises in correlated and topological systems. A detailed quantitative study of heavy fermion superconductivity in CeCoIn$_5$ was developed on the basis of cutting-edge scanning tunneling spectroscopy experiments [1, 2]. The low-energy heavy quasiparticle band structure and magnetic f-electron interaction were extracted from the quasiparticle interference data and used to calculate important properties of the superconducting state, including the gap symmetry and momentum dependence, the critical temperature, the spin-lattice relaxation rate, and the resonance peak observed in neutron scattering experiments. The strong agreement between experiment and theory demonstrates that a quantitative understanding of heavy fermion superconductivity is achievable in practice. This bolsters the case for similar combined experimental/theoretical studies in the future.

The model was also used to explain the features of the differential conductance both in the normal and superconducting states. In the former case, the observation of a purported pseudogap was explained in terms of van Hove singularities due to the flatness of the hybridized quasiparticle bands. In the latter, the gap seen in the dI/dV was shown to reflect the presence of the multiple superconducting gaps in the system. The local changes of the dI/dV in response to defects were also calculated in the model, in good agreement with experiment [3]. Next, the calculations of the resonance peak were extended to model its dispersion away from the commensurate antiferromagnetic wavevector [4]. These suggested that the peak was due to a magnon arising from the nearby antiferromagnetic phase, unlike the spin exciton observed in cuprate superconductors. It was also shown that the observed splitting of the resonance into two peaks under applied magnetic field can be understood in terms of magnetic anisotropy in the system.

The nonequilibrium behavior of a model nanoscale heavy fermion system was explored using the Keldysh Green's function approach. The charge current flow in the presence of an applied voltage bias was found to be sensitive to the Kondo-screening correlations between conduction and localized f-electrons, both in clean systems and those with defects. Coupling to phonons introduces a finite mean free path which limits the spatial extent of the modifications of the current pattern due to defects. The self-consistent calculation of the hybridization in the presence of a finite bias reveals the overall suppression of the correlations with increasing bias. However, the spatial structure of the hybridization does not show a monotonic decrease throughout the system, but rather some sites experience an increase in their hybridization. This could potentially be observed in scanning tunneling spectroscopy experiments.

Current flow in 2D topological insulators was also studied using the Keldysh technique [5]. Breaking the time-reversal symmetry in the system by introducing magnetic defects, we show that the edge states can be used to generate highly spin-polarized currents and design tunable spin diodes. The results are robust against various perturbations of the model and also found in systems in which TIs are interfaced with disordered ferromagnets or antiferromagnets. As such, they may find application in the developing fields of spintronics and quantum computation.

References

1. M.P. Allan, F. Massee, D.K. Morr, J. Van Dyke, A.W. Rost, A.P. Mackenzie, C. Petrovic, J.C. Davis, Imaging Cooper pairing of heavy fermions in $CeCoIn_5$. Nat. Phys. **9**(8), 468–473 (2013)
2. J.S. Van Dyke, F. Massee, M.P. Allan, J.C.S. Davis, C. Petrovic, D.K. Morr, Direct evidence for a magnetic f-electron–mediated pairing mechanism of heavy-fermion superconductivity in $CeCoIn_5$. Proc. Natl. Acad. Sci. **111**(32), 11663–11667 (2014)
3. J.S. Van Dyke, J.C.S. Davis, D.K. Morr, Differential conductance and defect states in the heavy-fermion superconductor $CeCoIn_5$. Phys. Rev. B **93**(4), 041107 (2016)
4. Y. Song, J.V. Dyke, I.K. Lum, B.D. White, S. Jang, D. Yazici, L. Shu, A. Schneidewind, P. Čermák, Y. Qiu, M.B. Maple, D.K. Morr, P. Dai, Robust upward dispersion of the neutron spin resonance in the heavy fermion superconductor $Ce_{1-x}Yb_xCoIn_5$. Nat. Commun. **7**, 12774 (2016)
5. J.S. Van Dyke, D.K. Morr, Controlling the flow of spin and charge in nanoscopic topological insulators. Phys. Rev. B **93**(8), 081401 (2016)

Appendix A
Keldysh Formalism for Transport

Various formalisms exist for the calculation of charge or spin currents in solids, ranging from semi-classical approaches to fully quantum mechanical ones [1]. Here the Keldysh Green's function method is used to determine the current flow in real space [2, 3]. Within this approach, the spin-resolved current between two sites is

$$I_{\mathbf{rr}'}^{\alpha} = -2\frac{e}{\hbar}\int_{-\infty}^{\infty}\frac{d\omega}{2\pi}\text{Re}[t_{\mathbf{rr}'}^{\alpha}G_{\alpha}^{<}(\mathbf{r},\mathbf{r}',\omega)] \tag{A.1}$$

where $\alpha = \uparrow, \downarrow$, $t_{\mathbf{rr}'}^{\alpha}$ is the electron hopping connecting the two sites (nearest or next-nearest), and $G_{\alpha}^{<}(\mathbf{r},\mathbf{r}',\omega)$ is the non-local dressed lesser Green's function. This is the Fourier transform to frequency space of the time domain lesser Green's function defined by $G_{\alpha}^{<}(\mathbf{r},\mathbf{r}',t,t) = \langle c_{\mathbf{r}'}^{\dagger}(t)c_{\mathbf{r}}(t)\rangle$. The total charge current through the system is thus $I_{\text{out}}^{c} = \sum_{\alpha} I_{\text{out}}^{\alpha}$ and the spin-α polarization of the outgoing current is given by $\eta_{\alpha} = I_{\text{out}}^{\alpha}/I_{\text{out}}^{c}$. One induces a current through the system by applying a chemical potential difference, $\mu_{L,R} = \pm eV/2$, in the left (L) and right (R) leads (corresponding to a voltage bias V across the sample). $G_{\alpha}^{<}(\mathbf{r},\mathbf{r}',\omega)$ is determined via the following Dyson equations

$$\hat{G}^{<} = \hat{G}^{r}\left[\left(\hat{g}^{r}\right)^{-1}\hat{g}^{<}\left(\hat{g}^{a}\right)^{-1} + \hat{\Sigma}_{ph}^{<}\right]\hat{G}^{a} \tag{A.2}$$

$$\hat{G}^{r} = \hat{g}^{r} + \hat{g}^{r}\left[\hat{t} + \hat{\Sigma}_{ph}^{r}\right]\hat{G}^{r} \tag{A.3}$$

Here $\hat{G}^{r}$ and $\hat{G}^{a}$ are the dressed retarded and advanced Green's functions, respectively. These arise from the non-interacting retarded ($\hat{g}^{r}$) and advanced ($\hat{g}^{a}$) Green's functions, which represent a lattice of completely decoupled sites. Explicitly,

$$\hat{g}^{r,a} = \begin{pmatrix} \hat{g}_{leads}^{r,a} & 0 \\ 0 & \hat{g}_{sys}^{r,a} \end{pmatrix} \tag{A.4}$$

© Springer International Publishing AG, part of Springer Nature 2018
J. S. Van Dyke, *Electronic and Magnetic Excitations in Correlated and Topological Materials*, Springer Theses, https://doi.org/10.1007/978-3-319-89938-1

where $\hat{g}^r_{sys}$ is the diagonal matrix with elements

$$g^r_{sys}(\omega) = \frac{1}{\omega + i\delta - eV_g} \tag{A.5}$$

and $\hat{g}^a_{sys} = (\hat{g}^r_{sys})^*$. Here e is the electron charge and V_g the gate voltage applied to select states at energy $E = eV_g$ for transport. On the other hand, the metallic leads are modeled using a constant density of states equal to unity,

$$g^r_{leads}(\omega) = -i\pi \tag{A.6}$$

The matrix $\hat{t}$ contains all the hopping elements connecting the various sites in the system and the system to the leads. Depending on the particular case under consideration, it will also include the effects of nonmagnetic and magnetic defects. The matrix $\hat{\Sigma}^r_{ph}$ describes the coupling of the system to phonons (discussed in more detail below). The non-interacting lesser Green's function has a similar form to the retarded and advanced ones:

$$\hat{g}^< = \begin{pmatrix} \hat{g}^<_{leads} & 0 \\ 0 & \hat{g}^<_{sys} \end{pmatrix} \tag{A.7}$$

Its components are again diagonal matrices, now with elements given by

$$g^<_{sys}(\omega) = -2in_F(\omega)\text{Im}g^r_{sys}(\omega) \tag{A.8}$$

$$g^<_{leads}(\omega) = -2in_F(\omega + \mu_{L,R}) \text{Im}g^r_{leads}(\omega) \tag{A.9}$$

To study the effect of coupling the system sites to local phonon modes, one introduces an electron-phonon term into the Hamiltonian

$$H_{e-ph} = g\sum_{\mathbf{r},\sigma} c^\dagger_{\mathbf{r},\sigma} c_{\mathbf{r},\sigma} \left(a^\dagger_{\mathbf{r}} + a_{\mathbf{r}}\right) + \sum_{\mathbf{r}} \omega_0 a^\dagger_{\mathbf{r}} a_{\mathbf{r}}, \tag{A.10}$$

Here g is the electron-phonon coupling strength, $a^\dagger_{\mathbf{r}}$ ($a_{\mathbf{r}}$) creates (annihilates) a phonon at site $\mathbf{r}$ in the system, and ω_0 is the phonon frequency. The summation runs over whatever sites are connected to local phonon modes. To simplify the calculation of the part of the electron self-energy arising from phonon interactions, we consider the high-temperature approximation in which $k_b T \gg \omega_0$ [4]. In this approximation one has $n_B(\omega_0) \gg 1$ so that only terms in $\hat{\Sigma}^r_{ph}$ containing $n_B(\omega_0)$ are kept. In the self-consistent Born approximation (i.e. using the dressed Green's function) one has

$$\Sigma^{r,<}_{\mathbf{rr}}(\omega) = ig^2 \int \frac{d\nu}{2\pi} D^<(\nu) G^{r,<}_{\mathbf{rr}}(\omega - \nu), \tag{A.11}$$

A Keldysh Formalism for Transport

with phonon Green's functions

$$D_0^<(\omega) = 2i n_B(\omega) \mathrm{Im} D_0^r(\omega) \tag{A.12}$$

$$D_0^r(\omega) = \frac{1}{\omega - \omega_0 + i\delta} - \frac{1}{\omega + \omega_0 + i\delta} \tag{A.13}$$

which are assumed to not be modified by the applied bias. To obtain an analytical expression for $\Sigma_{\mathbf{rr}}^{r,<}(\omega)$ it is useful to further take the limit $\omega_0 \to 0$, in which case one has to leading order in $k_b T/\omega_0$ that

$$\Sigma_{\mathbf{rr}}^{r,<}(\omega) = 2g^2 \frac{k_B T}{\omega_0} G_{\mathbf{rr}}^{r,<}(\omega) \equiv \gamma G_{\mathbf{rr}}^{r,<}(\omega) \tag{A.14}$$

This is conveniently expressed by the introduction of a superoperator $\tilde{D}$, which acting on a matrix returns a new matrix with all elements equal to zero except those on the diagonal assigned to sites coupled to phonon modes,

$$\left[\tilde{D}\hat{G}^{r,<}\right]_{\mathbf{rr'}} = \begin{cases} G_{\mathbf{rr'}}^{r,<} \delta_{\mathbf{rr'}} & \text{if } g \neq 0 \text{ at site } \mathbf{r} \\ 0 & \text{otherwise} \end{cases} \tag{A.15}$$

With this notation one may write

$$\Sigma^{r,<}(\omega) = \gamma \tilde{D} \hat{G}^{r,<} \tag{A.16}$$

It is also useful to define the superoperator $\hat{U}$ acting on a matrix $\hat{X}$ by

$$\hat{U}\hat{X} = \hat{G}^r \hat{X} \hat{G}^a \tag{A.17}$$

Together these superoperators permit the solutions of the Dyson equations to be concisely written as

$$\hat{G}^< = \hat{U} \left[1 - \gamma \tilde{D} \hat{U}\right]^{-1} \hat{\Lambda} \tag{A.18}$$

$$\hat{G}^r = \left[1 - \hat{g}^r \left(\hat{t} + \gamma \tilde{D} \hat{G}^r\right)\right]^{-1} \hat{g}^r \tag{A.19}$$

where $\hat{\Lambda} = \hat{g}_r^{-1} \hat{g}^< \hat{g}_a^{-1}$ is a diagonal matrix. The only non-zero elements of $\hat{\Lambda}_{\mathbf{rr}}$ are those for which $\mathbf{r}$ is a site in one of the leads. Expanding Eq. (A.18) yields

$$\hat{G}_{\mathbf{rr'}}^< = \sum_{\mathbf{l}} \hat{G}_{\mathbf{rl}}^r \left[\hat{\Lambda}_{\mathbf{ll}} + \gamma \sum_{\mathbf{m}} \hat{Q}_{\mathbf{lm}} \hat{\Lambda}_{\mathbf{mm}} + \gamma^2 \sum_{\mathbf{m,p}} \hat{Q}_{\mathbf{lm}} \hat{Q}_{\mathbf{mp}} \hat{\Lambda}_{\mathbf{pp}} + \ldots \right] \hat{G}_{\mathbf{lr'}}^a$$

$$\tag{A.20}$$

with

$$\hat{Q}_{lm} = \begin{cases} |G^r_{lm}|^2 & \text{if } g \neq 0 \text{ at site } l \\ 0 & \text{otherwise} \end{cases} \qquad (A.21)$$

This may be further simplified by defining the vector $\lambda_l \equiv \hat{\Lambda}_{ll}$ so that

$$\hat{G}^<_{rr'} = \sum_l \hat{G}^r_{rl} \left[\left(1 - \gamma \hat{Q}\right)^{-1} \lambda \right]_l \hat{G}^a_{lr'} \qquad (A.22)$$

Now defining the diagonal matrix

$$\tilde{\Sigma}_{ll} = \left[\left(1 - \gamma \hat{Q}\right)^{-1} \lambda \right]_l \qquad (A.23)$$

the final expression for the lesser Green's function in the presence of phonons takes the simple form

$$\hat{G}^< = \hat{G}^r \tilde{\Sigma} \hat{G}^a \qquad (A.24)$$

References

1. C. Jacoboni, in *Theory of Electron Transport in Semiconductors. Springer Series in Solid-State Sciences*, vol. 165 (Springer, Berlin, 2010)
2. L.V. Keldysh, Diagram technique for nonequilibrium processes. Zh. Eksp. Teor. Fiz. **47**, 1515 (1964). [Sov. Phys. J. Exp. Theor. Phys. **20**, 1018 (1965)]
3. C. Caroli, R. Combescot, P. Nozieres, D. Saint-James, Direct calculation of the tunneling current. J. Phys. C Solid State Phys. **4**(8), 916 (1971)
4. Z. Bihary, M.A. Ratner, Dephasing effects in molecular junction conduction: an analytical treatment. Phys. Rev. B **72**(11), 115439 (2005)

CPSIA information can be obtained
at www.ICGtesting.com
Printed in the USA
LVHW07*2016240518
578390LV00001B/1/P